STUDENT'S SOLU for Hildebran

Statistical Thinking For Managers

Fourth Edition

David K. Hildebrand
Department of Statistics
Wharton School
University of Pennsylvania

Patricia Hildebrand
School of Arts and Sciences
Computing
University of Pennsylvania

Assisted by

Jeffrey D. Hildebrand
Department of Mathematics
University of Wisconsin

Stephanie Cole
Merck Research Laboraties

DUXBURY PRESS
An Imprint of Brooks/Cole Publishing Company
I(T)P® An International Thomson Publishing Company

Pacific Grove • Albany • Belmont • Bonn • Boston • Cincinnati • Detroit
Johannesburg • London • Madrid • Melbourne • Mexico City • New York
Paris • Singapore • Tokyo • Toronto • Washington

Assistant Editor: *Cynthia Mazow*
Editorial Assistant: *Rita Jaramillo*
Marketing Representative: *Laura Hubrich*
Production Editor: *Mary Vezilich*

I(**T**)**P** The ITP logo is a registered trademark used herein under license.
Duxbury Press and the leaf logo are trademarks used herein under license.

*For more information, contact Duxbury Press at Brooks/Cole Publishing Company,
or electronically at* **http://www.duxbury.com**

BROOKS/COLE PUBLISHING COMPANY
511 Forest Lodge Road
Pacific Grove, CA 93950
USA

International Thomson Publishing Europe
Berkshire House 168-173
High Holborn
London WC1V 7AA
England

Thomas Nelson Australia
102 Dodds Street
South Melbourne, 3205
Victoria, Australia

Nelson Canada
1120 Birchmount Road
Scarborough, Ontario
Canada M1K 5G4

International Thomson Editores
Seneca 53
Col. Polanco
11560 México, D. F., México

International Thomson Publishing Japan
Hirakawacho Kyowa Building, 3F
2-2-1 Hirakawacho
Chiyoda-ku, Tokyo 102
Japan

International Thomson Publishing Asia
60 Albert Street
#15-01 Albert Complex
Singapore 189969

International Thomson Publishing GmbH
Königswinterer Strasse 418
53227 Bonn
Germany

Printed in Canada

5   4   3   2   1

ISBN 0-534-35372-x

Table of Contents

Chapter 2

Summarizing Data about One Variable

2.1   The Distribution of Values of a Variable

2.2   The task here is to interpret the histogram, rather than to construct it. The Statistix package has done the basic work.

a. The average value falls in the middle of the histogram, where the histogram balances. This particular histogram is close to symmetric, with nearly equal left and right tails. It appears to us that the histogram would balance somewhere in the interval between 842 and 847. Therefore the average value should be about 845.
   Variability is a matter of how spread out the histogram is. In this case, there is a value in the 812 to 817 interval, and, on the other side, a value in the 867 to 872 interval. (Note in both intervals, the frequency is shown as 1, so there is only one value.) We'd say that was a fair amount of variability.
   Skewness is the opposite of symmetry. In this case, the two tails are just about equal, and it would be hard to say which direction had the longer tail. We'd say there was virtually no skewness.

b. The smoothed component is shown as a curve through the histogram. Skewness is indicated by a nonsymmetric curve. In this case, the curve appears almost perfectly symmetric around the middle peak. There is little or no skewness.

2.3   Recall that a stem-and-leaf display, like the one shown, groups the data according to the values in the stem. The first value shown must be 812 (as opposed to 81.2 or 8120), from a look at the data.
   The stem-and-leaf display gives interval widths of 10, in contrast to the width of 5 in the histogram. In effect, the stem-and-leaf display centers the intervals at 815, 825, etc.; the centers for the histogram also differ. The two pictures aren't identical.
   We see basically the same pattern in both displays. The average value is somewhere in the 840's, there is modest variability, and there is very little skewness.

2.6   Again, our task is interpretation of output. The H and M notations need not bother us; we should use the part of the display that we understand.

a. Recall that skewness is a nonsymmetric pattern in the plot. There appears to be a long tail extending toward larger values. Most of the numbers are in the teens, but a few are as large as the thirties and forties. (Of course, there are no negative values.) In a histogram, the long tail would extend toward larger values on the right, the large values. Thus the data appear right skewed.

b. To construct a histogram, we need to define classes, count the number of values falling in each class, and draw rectangles corresponding to each class.

First we must choose convenient classes. The data range from 7.7 to 45.9, a range of $45.9 - 7.7 = 38.2$ points. If we chose a class width of 10, there would only be about 4 classes. A width of 5 would be better, yielding about 8 classes. One of many ways to have class width 5 is to center the intervals at 10, 15, ..., 45. Assuming we use these midpoints, the classes are 7.50–12.49, 12.50–17.49, and so on. We wouldn't have any values that fall right on the class endpoints (such as 12.50) for these data, but it's better to be unambiguous in defining the classes.

A frequency table using these classes, obtained simply by counting the number in each class, follows:

Class	Midpoint	Frequency
7.50-12.49	10	7
12.50-17.49	15	5
17.50-22.49	20	4
22.50-27.49	25	1
27.50-32.49	30	2
32.50-37.49	35	3
37.50-42.49	40	1
42.50-47.49	45	1
Total		24

The relative frequencies are the frequencies divided by 24, the number of measurements. The resulting histogram follows.

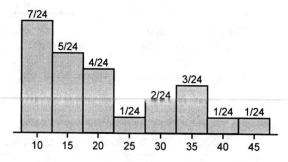

❋ *Note:* Error in the textbook. The first two columns of the Excel spreadsheet are cut off; the missing numbers are 7.7, 8.5, 14.2, 13.7.

2

2.2   On the Average: Typical Values

2.10   The mean and median are two measures of "average" value. They differ when there is skewness. In this case, the mean is larger than the median, 1090.65 vs. 1039. This fact indicates that the mean is being pulled to the right by some large values. Therefore, there should be right skewness in the data. It's not so evident in the histogram that the skewness is in that direction.

2.3   Measuring Variability

2.13   **a.** The range is defined to be the difference between the largest and smallest values. Remember that it is affected by outliers and by sample size. The range of the observations in sample 1 is therefore

$$25 - 15 = 10$$

The range for sample 2 is

$$26 - 14 = 12$$

b. The data are stated as samples, so we should use the sample standard deviation. The definition says that we should first calculate the mean, then deviations from the mean. Square these, add the squared deviations, and divide by $n - 1$. Finally, take the square root.

Sample 1

	y_i	$(y_i - \bar{y})$	$(y_i - \bar{y})^2$
	15	−5	25
	19	−1	1
	21	1	1
	25	5	25
Total	80		52

$$\bar{y} = \frac{80}{4} = 20$$

$$s^2 = \frac{52}{3} = 17.33$$

$$s = 4.16$$

Sample 2

y_i	$(y_i - \bar{y})$	$(y_i - \bar{y})^2$
14	−6	36
17	−3	9
18	−2	4
19	−1	1
19	−1	1
20	0	0
20	0	0
20	0	0
21	1	1
21	1	1
22	2	4
23	3	9
26	6	36
Total	260	102

$$\bar{y} = \frac{260}{13} = 20$$

$$s^2 = \frac{102}{12} = 8.5$$

$$s = 2.92$$

c. Comparing the standard deviations for the two samples, sample 1 shows more variability. This is supported by comparison of boxplots for both samples. Remember that greater variability appears as a wider boxplot.

For example, here are two boxplots created by Minitab, version 10. The box for sample 1 is much wider, indicating greater variability. True, the whiskers for sample 2 extend a bit farther; still, the narrower box for sample 2 strongly suggests that sample 1 is more variable.

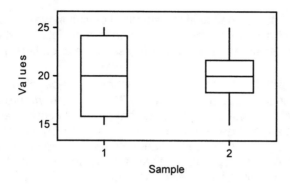

Note that the ranges computed in part **a.** don't reflect the greater variability in sample 1, because of the larger sample size in sample 2.

2.16   There will be many sources of variability, and you can probably think of different ones than we can. Operators will differ in experience, skill, and possibly training, all of which will lead to variability in number of calls cleared. The nature of calls will vary. Some will be more complicated than others. Depending on which operator gets which calls, differences in the nature of calls will also be a source of variability in number of calls cleared. There may be other sources such as differences in equipment or even differences in operators' speaking speed.

2.17   **a.** The mean is shown as `794.23`; the standard deviation (`STDEV`) is shown as `34.25`. Therefore, mean minus one standard deviation is $794.23 - 34.25 = 759.98$ and mean plus one standard deviation is $794.23 + 34.25 = 828.48$. The actual data are whole numbers, not decimals; all values between 760 and 828 will fall in this interval.

b. 51 out of 60 is 85%; $51/60 = 0.85$. According to the Empirical Rule, the percentage theoretically should be only 68%. The one standard deviation interval is too wide in this case. It seems likely that skewness or outliers have inflated the standard deviation. This will make the interval "too wide" and capture "too many" of the data values.

2.18   Recall that outliers are shown in a boxplot as points beyond the "whiskers" of the plot. The boxplot shows several outliers, including one very serious one. These outliers will inflate the standard deviation, making the " one standard deviation interval wide and causing the Empirical Rule to fail.

2.19   **a.** The question does not ask for calculations, but interpretation. Note that the 601 score is an outlier, on the low end of the scale.
    Deleting the lowest score should increase the average (mean) somewhat; it was pulling down the mean. Deleting any outlier (whether low or high) should decrease the standard deviation; the remaining data will appear less variable.

b. The mean increased, as expected, from 794.23 to 797.51. The magnitude of change is fairly small. The standard deviation decreased, as expected, from 34.25 to

23.21. Notice that the change in the standard deviation was much more dramatic than the small change in the mean. The standard deviation and variance are based on squared deviations and are much more sensitive to outliers than is the mean.

2.27   **a.** A problem with the mean would show in the top, x-bar chart. The mean chart shows control limits at about 16.67 and 15.33. (These numbers appear on the right side of this particular chart.) The values that are out of control are indicated by a * symbol. There are a large number of * symbols, indicating a serious problem with the mean values. Generally, there is a clear downward trend in the means, as opposed to a sudden drop. It is true that there was a large decrease at day 20, but there is also an evident downward trend elsewhere.

b. Variability should be reflected in the range chart, the second of the control charts. None of the range numbers is even close to the upper control limit, shown on the right of this particular chart as about 2.46. There is no evidence of any kind of problem of excess variability.

2.28   This exercise asks for an interpretation of the results rather than a calculation. The two explanations would lead to different types of out of control patterns. If bad lots were a problem, there ought to be noticeable jumps when a new lot is used. If the problem is with the heating system, there should be more of a drift, rather than a jump, because the change should be gradual. The control chart seems to indicate a decreasing trend in the means, rather than a sudden jump. There doesn't seem to be any identifiable point (except perhaps day 20) at which the means change suddenly. This fact suggests that the problem is a gradual worsening of the heating system.

2.29   **a.** The skewness is revealed by the points falling off the line — the points are steeper than the line.

b. The outlier is shown as the single point in the lower left corner of the plot (at the point (600, 0.01)). If this data were truly normal, we would expect a value of 600 to occur less frequently (corresponding to a probability smaller than 0.01 — one that falls on the line).

2.31   Overall, these data are close to normally distributed. The only deviation occurs in the extreme tails (shown by the deviation from the line at the ends).

Supplementary Exercises

2.36   **a.** To construct a histogram, decide on how to divide the data into classes, count the number of measurements in each class, and construct rectangles with these counts as heights. With 25 observations, 5 classes seem suitable. The values go from 8 to 32, a range of 24; therefore, try a class width of 5.

Class	Midpoint	Frequency
7.5–12.4	10	10
12.5–17.4	15	0
17.5–22.4	20	5
22.5–27.4	25	0
27.5–32.4	30	10

The histogram of the data as grouped in the frequency table is displayed below:

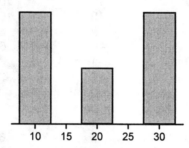

The data are symmetric (no skewness) and trimodal. That is, there are three separate, distinct peaks in the histogram.

b. Remember that the mean and median are measures of the middle of the data. In this artificial case, the data are distributed *perfectly* symmetrically about the value 20 (see the histogram in part **a.**). Therefore the mean and median are exactly the same, 20.

c. Recall that the standard deviation is the square root of the variance. In turn, the variance is the sum of squared deviations around the mean, divided by the sample size minus 1.

$$s^2 = \frac{1}{n-1}\left(\sum_{i-1}^{n}(y_i - \bar{y})^2\right)$$

Computation of s^2 is facilitated by the use of columns of computation as set up below:

y_i	$(y_i - \bar{y})$	$(y_i - \bar{y})^2$
8	−12	144
9	−11	121
10	−10	100
10	−10	100
10	−10	100
10	−10	100
10	−10	100
10	−10	100
11	−9	81
12	−8	64
19	−1	1
20	0	0
20	0	0
20	0	0
21	1	1
28	8	64
29	9	81
30	10	100
30	10	100
30	10	100
30	10	100
30	10	100
30	10	100
31	11	121
32	12	144
Total		2,022

$$s^2 = \frac{1}{24}(2{,}022) = 84.25$$

Therefore,

$$s = \sqrt{84.25} = 9.18$$

d. The interval within 1 s of the mean is 20 ± 9.18 or

10.82 to 29.18

In this case the data are integers, so anything from 11 through 29 will be included in the interval. The number of observations that actually fall within this interval is 9 or *36%*. Applying the Empirical Rule, this interval should contain approximately *68%* of the observations which is considerably more than actually occurred.

8

The Empirical Rule holds when the values have roughly a mound-shaped histogram. The data in this case don't have this type of histogram, but instead have three different modes. That's why there is such a discrepancy in the fraction of values that falls within 1 s of the mean.

e. The interval within 2 s of the mean is $20 \pm 2(9.18)$ or

$$1.64 \text{ to } 38.36$$

The number of observations that actually fall within this interval is all 25 or *100%*. This is more than the 95% indicated by the Empirical Rule. The two-standard-deviation interval picked up all three "humps" of data, as it happens.

2.42 We are looking for some place where the data (mean lateness) changes noticeably. The means for the last five days seem noticeably larger. Up until that point, the means all are roughly 1 or 2 minutes. Then, suddenly the means jump to 4 minutes, more or less.

2.43 We are looking for a place where the data (means) indicate a problem. In this context, a problem would imply a high value for average lateness. Four of the five means at the end of the data (day number 53 and beyond) definitely are higher than previous means. There are no other evident places where there is a dramatic shift.

2.44 **a.** Recall that the control limits are defined as the target mean plus and minus three standard deviations. The target mean is 1.5 and the standard deviation is 0.30. Three standard deviations is $3(0.30) = 0.90$. The control limits are $1.5 + 0.90 = 2.40$ and $1.5 - 0.90 = 0.60$.

b. Look at the list of means or at the control chart. There are no means that fall below the lower control limit, 0.60. (And if there were, that would be fine; a low amount of lateness is desired.) The means for days 22, 30, and 32 fall slightly above the upper control limit; the means for days 54, 55, 57 and 58 fall far above the upper control limit and indicate a definite problem.

2.45 Variability is measured by standard deviations, so we should look to the S chart. The largest standard deviations all occur right at the end of the data, for days 54 through 58. Except for day 56, these are exactly the days for which the means are highest. The high variability days do coincide with the high mean days.

2.53 This requires some thought, and is something of a trick question. Normally, the fact that a mean is larger than a median indicates right skewness. However, in this case, the numbers are not quantitative measurements, but only qualitative codes. They aren't even ordered in any evident way. For qualitative, unordered data, the mean and median are meaningless. Therefore, the finding that the mean is larger than the median doesn't indicate anything relevant.

2.54   **a.** Recall that a Pareto chart is basically a histogram, with frequencies (counts) indicated by the heights of the rectangles. Codes 6, 9, and 7 have much the highest frequencies, as shown in the Pareto chart. We would say that these three codes accounted for the great majority of reasons for leaving.

b. Again we must consider the context. Look at the definitions of the codes. These three codes all have to do with pay.

2.55   The combined codes (and the labels below the bars) rather clearly indicates that Pay is the dominant reason for leaving.

2.61   **a.** Here we want to think, not calculate. The mean is a measure of the middle point of the data, and the value where a histogram would balance. There are many scores in the 20's, 30's and 40's, but a few scores that are much larger, up in the hundreds. The large values will pull up the mean. Therefore, we'd guess that the mean is about 50 or 60.

b. Again, where is the middle of the data? Most BIDPERHR values begin with 0 or 1, but there are some 2 and 3 numbers. Therefore, we'd guess that the mean is about 1.0 or so.

2.62   **a.** The mean is where each histogram balances. The skewness in the MINPRBID histogram suggests that the mean might be somewhat larger than 50. With such severe skew, it's hard to make a good guess. In the BIDPERHR histogram, the mean seems to be about 1.0, perhaps bigger, but not as large as 2.0.

b. The MINPRBID histogram is severely right-skewed. The BIDPERHR histogram is much less skewed.

2.63   **a.** No calculations are needed. The respective means are shown as 62.462 and 1.432. Note that we guessed a bit low in previous answers. You may have done better.

b. The reason for the question is that BIDPERHR is defined as 60/MINPRBID. Do the mean values work the same way? No; the value of 60/mean(MINPRBID), 60/62.462, is less than 1.0 and quite different from the mean for BIDPERHR, 1.432. Therefore, the means do not have the same relation as the variables themselves.

Chapter 3

A First Look at Probability

3.1   Basic Principles of Probability

3.3   For this exercise, we are choosing randomly from a specified set. The classical interpretation applies; all we have to do is understand what question is being asked, and count to find the answer.

a. We are looking for the unconditional probability of brand C. We are given that 541 of the total of 723 problems come from brand C. Therefore, by the classical interpretation of probability,

$$P(C) = \frac{541}{723} = 0.748$$

b. We want the probability of problems with the engine *or* the transmission. These are regarded as mutually exclusive possibilities in the table. There are a total of $127 + 326 = 453$ problems involving either the engine or the transmission, out of the total of 723.

$$P(\text{Engine or Transmission}) = \frac{453}{723} = 0.627$$

c. The problems for which the dealer is *not* fully reimbursed are exhaust and other problems for brand G. There are $16 + 6 = 22$ such problems, so

$$P(\text{Dealer not fully reimbursed}) = \frac{22}{723} = 0.030$$

3.4   The important thing to understand in this exercise is that we need conditional probabilities. The conditioning event (the given event) is the brand. The random event is the kind of problem.

a. The relevant problems are only the 541 problems for brand C. Of these, 106 are engine problems. Therefore

$$P(\text{Engine}|C) = \frac{106}{541} = 0.196$$

b. We may construct a table by the same sort of calculations used in answering part **a.** Given a particular brand, we want to divide the frequency for that brand and type of problem by the total frequency for that brand.

		Engine	Transm.	Exhaust	Fit/finish	Other
Given brand	C	0.196	0.390	0.124	0.246	0.044
	G	0.115	0.632	0.088	0.132	0.033

(Column header above: Problem area)

The probabilities aren't entirely similar. Given G, the probability of a transmission problem is substantially higher. Given C, the probabilities of all the other problems are higher.

3.5   **a.** Of the total of 609 jobs, $70 + 22 = 92$ involve more than one problem (that is, two or three problems). So

$$P(\text{more than one problem}) = \frac{92}{609} = 0.151$$

b. This is a conditional probability. Only the brand C information is relevant. Given brand C, there are 453 jobs, of which $54 + 17 = 71$ involve more than one problem.

$$P(\text{more than one problem}|C) = \frac{71}{453} = 0.157$$

3.6   The table may be constructed using the same sort of calculation as in part **b.** of the previous exercise. Be careful to consider the correct condition.

		1	2	3
Given brand	C	0.843	0.119	0.038
	G	0.865	0.103	0.032

(Column header above: Number of problems)

The conditional probabilities are nearly identical.

3.7   This question requires you to think, not calculate. If the number of cars sold were the same for the two brands, the higher number in brand C would indicate quality problems. But we don't have any information about how many of the two brands were sold, so we can't compare them as to quality.

3.2 Statistical Independence

3.16   Let

A be the event "no substitute needed at primary school 1,"
B be the event "no substitute needed at primary school 2," and
C be the event "no substitute needed at the high school"

The information in the problem indicates that

$$P(A) = 0.60 \quad P(B) = 0.60 \qquad P(C) = 0.50$$

We are assuming statistical independence.
   To find the probability that no substitute will be needed at any of the schools, find $P(A \text{ and } B \text{ and } C)$.

$$
\begin{aligned}
P(A \text{ and } B \text{ and } C) &= P(A)P(B)P(C) \\
&= 0.60(0.60)(0.50) \\
&= 0.18
\end{aligned}
$$

Note that the independence assumption means that we do not need to use conditional probabilities.

3.17   The assumption of independent processes is not a realistic one, we think. The reasons why substitutes are needed in one school would most likely be reasons for needing substitutes in the other schools, as well. Take the case of a flu epidemic which affects an entire region. If substitutes are needed at primary school 1 because teachers are ill, most likely substitutes will be needed at primary school 2 and at the high school because of teacher absence. Further, if bad weather leads to absences in one school, it will lead to absences in other schools in the same area.

3.21   Just as in the previous exercise, the 4% figure was derived by assuming that the two events are statistically independent. We know of no reason why telephone orders should be either more likely or less likely to lead to returns than other types of orders. Therefore, we would think it quite plausible that the events are at least close to independent and that the 4% figure is at least close to correct.

3.22   Here the computer has done the "heavy lifting" for us, and we need to interpret the results. The program looks at the entire 60-day period and counts the number of times a run of at least 8 has occurred. The long-run relative frequency interpretation of probability applies here, as long as we agree that 10,000 trials is an adequately long run.

The output indicates that the probability of obtaining *no* runs of at least 8, in a 60-day period, is about 0.808 (80.8%). By the complements principle, the probability of obtaining at least one run is $1 - 0.808 = 0.192$. This is a surprisingly large probability, to most people.

3.23 A major problem with finding the probability of at least one run is that the run could occur anywhere within the 60-day period. We could perhaps find the probability of a run of eight in the first eight days, but that is not what's asked. We are trying to find the probability of a run of at least 8 *somewhere* in the 60-day period.
Probably the best answer to the question is "with great difficulty!"

3.3   Probability Tables, Trees, and Simulations

3.27 Once again, a good way to start is by writing down the relevant probability information. The following probabilities are given:

$$P(\text{regular bidder}) = 0.60$$
$$P(\text{occasional bidder}) = 0.30$$
$$P(\text{first time bidder}) = 0.10$$
$$P(\text{satisfactory service}|\text{regular bidder}) = 0.90$$
$$P(\text{satisfactory service}|\text{occasional bidder}) = 0.80$$
$$P(\text{satisfactory service}|\text{first time bidder}) = 0.60$$

With this information, we could proceed in (at least) two different ways. We could construct a table, with type of bidder as the rows and quality of service (satisfactory or unsatisfactory) as the columns. Alternatively, we could construct a probability tree. To build the tree, we would start with three branches for type of bidder, because we have unconditional probabilities of the three types. Then we would add a second set of branches for quality of service, because we have conditional probabilities of these. Then insert the probabilities and multiply to obtain path probabilities.

A probability tree is constructed below:

path

		Satisfactory	0.90	$0.60(0.90) = 0.54$	1
Regular Bidder	0.60				
		Unsatisfactory	0.10	$0.60(0.10) = 0.06$	2
		Satisfactory	0.80	$0.30(0.80) = 0.24$	3
Occasional Bidder	0.30				
		Unsatisfactory	0.20	$0.30(0.20) = 0.06$	4
		Satisfactory	0.60	$0.10(0.60) = 0.06$	5
First Time Bidder	0.10				
		Unsatisfactory	0.40	$0.10(0.40) = 0.04$	6

a. The probability may be read directly from the tree.

$$P(\text{first time bidder and satisfactory service}) = 0.06$$
(Path 5 on probability tree)

b. This question also can be answered directly from the tree. Add Paths 1, 3, and 5 on probability tree: $0.54 + 0.24 + 0.06 = 0.84$. Therefore, $P(\text{satisfactory service}) = 0.84$

c. This question asks for a conditional probability. One sensible approach is to try using the definition of conditional probability. $P(\text{first time bidder|satisfactory service}) =$

$$\frac{P(\text{first time bidder and satisfactory service})}{P(\text{satisfactory service})} = \frac{0.06}{0.84} = 0.071$$

3.32 The output indicates that 3 individuals both had a positive diagnosis and had the disease. Notice in lines 29 and 30 of the output, TOTBOTH is the number with both positive diagnosis and the disease. The total number of positive diagnoses was 100. In lines 27 and 28, the program counted how many people had a positive diagnosis. The fraction shown, 0.03, is therefore the approximate probability of having the disease, given a positive diagnosis.

3.33 As usual, we start branching in a probability tree with unconditional probabilities. In this situation, the unconditional probabilities refer to whether or not the person has the disease. At the second branch, we label positive or negative diagnoses. Then multiply to obtain the path probabilities, as shown in the following tree.

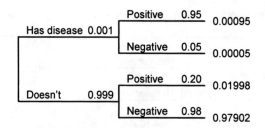

To find the probability of having the disease given a positive diagnosis, use the definition of conditional probability.

$$P(\text{Has disease}|\text{positive}) = \frac{P(\text{Has disease and positive})}{P(\text{positive})}$$

$$= \frac{0.00095}{0.00095 + 0.01998} = 0.045$$

Supplementary Exercises

3.35   **a.** We want P(MBA or undergraduate business degree). The "or" indicates that we should add probabilities. The two events are not mutually exclusive, so we must use the general form of the addition principle.

$$P(\text{MBA or undergrad}) = P(\text{MBA}) + P(\text{undergrad}) - P(\text{both})$$
$$= 0.29 + 0.24 - 0.08 = 0.45$$

Alternatively, we could notice that 21% of the VP's have an MBA but not an undergraduate business degree; these are the 29% who have an MBA less the 8% who have both. Similarly, 16% = 24% – 8% have an undergraduate business degree but not an MBA, while 8% have both. The event (MBA or undergrad) may be thought of as (MBA and not undergrad) or (undergrad and not MBA) or (both). Thought of this way, we have three mutually exclusive possibilities and may simply add probabilities

$$P(\text{MBA or undergrad}) = 0.21 + 0.16 + 0.08 = 0.45$$

once again

b. Holding neither degree is the complement of holding at least one degree. Therefore, we can use the complements principle.

$$P(\text{neither}) = 1 - P(\text{MBA or undergrad}) = 1 - 0.45 = 0.55$$

3.36 There are several ways to approach the problem. Perhaps the easiest is to note that 21% of VP's have an MBA but not an undergraduate business degree and that 16% have an undergraduate business degree but not an MBA. These are the two mutually exclusive ways of having one or the other degree, but not both.

$$P(\text{MBA or undergrad, but not both}) = 0.21 + 0.16 = 0.37$$

Alternatively, we know that $P(\text{MBA or undergrad, or both}) = 0.45$. We wish to exclude the 8% who hold both degrees, so

$$P(\text{MBA or undergrad, but not both}) = 0.45 - 0.08 = 0.37$$

3.46 **a.** Let the event F be "customer pays in full". We have $P(F) = 0.50$, and must assume that this probability applies to both months. With subscripts 1 and 2 to indicate the two customers, we note that the event "both customers pay in full" is $(F_1 \text{ and } F_2)$. Use multiplication to find this "and" probability.

$$P(F_1 \text{ and } F_2) = P(F_1)P(F_2) = (0.50)(0.50) = 0.25$$

b. Without additional information we must assume that $P(F \text{ in month 2} | F \text{ in month 1}) = P(F \text{ in month 2}) = 0.50$. Letting the subscripts indicate the month

$$P(F_1 \text{ and } F_2) = (0.50)(0.50) = 0.25$$

c. In both parts, we assumed that the events F_1 and F_2 are independent. In part **a.**, the assumption is reasonably sensible; the result for one customer should not affect the probabilities of another customer's behavior. In part **b.**, the assumption seems quite unreasonable. A customer who pays in full one month is presumably more likely to pay in full the next month.

3.47 **a.** As in the previous exercise, define F_1 to be "pays first month in full" and F_2 to be "pays second month in full". From that exercise, we have $P(F_1) = 0.50$. The information in this exercise is that $P(F_2 | F_1) = 0.90$ and $P(F_2 | \overline{F_1}) = 0.10$. Again we seek an "and" probability. Use the multiplication principle.

$$P(F_1 \text{ and } F_2) = P(F_1)P(F_2 | F_1) = (0.50)(0.90) = 0.45$$

b. Notice that this event is *not* the complement of the event in part **a.** (The customer could pay neither, pay both, or pay one of the two in full.) We want the probability of not $-F_1$ and not $-F_2$.

$$P(\overline{F}_1 \cap \overline{F}_2) = P(\overline{F}_1)P(\overline{F}_2|\overline{F}_1) = (1-0.50)(1-0.10) = 0.45$$

c. Note that the event "exactly one of two consecutive bills in full" is *not* the same as $(F_1 \text{ or } F_2)$ because F_1 or F_2 includes the possibility of both events. One way to answer this question is to break the event up into "pay in full month 1 and not month 2" or "pay in full month 2 and not month 1."

$$P(\text{exactly one month}) = P\left(\left[F_1 \text{ and } \overline{F}_2\right] \text{ or } \left[\overline{F}_1 \text{ and } F_2\right]\right)$$
$$= P(F_1 \text{ and } \overline{F}_2) + P(\overline{F}_1 \text{ and } F_2) \text{ by the addition law}$$
$$P(\text{exactly one month}) = P(F_1)P(\overline{F}_2|F_1) + P(\overline{F}_1)P(F_2|\overline{F}_1)$$
$$= (0.50)(1-0.90) + (1-0.50)(0.10) = 0.10$$

Alternatively, we note that there are three mutually exclusive possibilities: paying in full in both of two months, in neither of two months, or in exactly one of two months. Using parts **a.** and **b.**,

$$P(\text{exactly one month}) = 1 - P(\text{both}) - P(\text{neither})$$
$$= 1 - 0.45 - 0.45 = 0.10$$

3.48   We want to use the definition of conditional probability here.

$$P(F_1|F_2) = \frac{P(F_1 \cap F_2)}{P(F_2)}$$

In the previous exercise, we obtained $P(F_1 \text{ and } F_2) = 0.45$. One way to find $P(F_2)$ is to break it into cases, corresponding to whether or not the customer paid the first month's bill in full.

$$P(F_2) = P(F_1 \text{ and } F_2) + P(\overline{F}_1 \text{ and } F_2)$$
$$= P(F_1)P(F_2|F_1) + P(\overline{F}_1)P(F_2|\overline{F}_1)$$
$$= (0.50)(0.90) + (1-0.50)(0.10) = 0.50$$

Thus

$$P(F_1|F_2) = \frac{0.45}{0.50} = 0.90$$

Alternatively a tree or table could be used.

3.55 In the output, a note indicates that having the minimum value equal 1 means that the equipment works; a minimum equal 0 means that it fails. The output shows that the equipment worked in 610 of 1,000 trials. Because 1,000 trials is not an infinitely long run, we can't say we have exactly the correct probability, but 1,000 trials should make the probability quite close.

$$P(\text{equipment works}) \approx \frac{610}{1000} = 0.610$$

3.56 The equipment works if component 1 works and component 2 works and component 3 works, and so on all the way through component 100. The probability for any one component working is 0.99. We will have to assume that the components work or fail independently.

$$P(C_1 \text{ and } C_2 \text{ and } \dots C_{100}) = P(C_1)P(C_2)\cdots P(C_{100}) = (0.99)^{100}$$
$$= 0.605$$

This is very close to the simulation value, 0.610.

3.57 Just as before, a minimum equal to 1 means that the equipment works. In 1,000 trials each with a probability 0.995 of working, the equipment worked 604 times. Again, this is only a very good approximation, because we don't have an infinite set of trials.

$$P(\text{equipment works}) \approx \frac{604}{1,000} = 0.604$$

Review Exercises—Chapters 2 and 3

R.3 First, summarize the information. We are given $P(\text{increase}) = 0.60$, $P(\text{decrease}) = 0.40$, $P(\text{correct|increase}) = 0.93$, and $P(\text{correct|decrease}) = 0.98$. We want to find $P(\text{decrease|not correct})$.

There are several ways to do the problem. For illustration, we use a table, though a tree or basic reasoning could also be used.

	Correct	Not Correct	
Increase	0.60(0.93) = 0.558	0.042	0.60
Decrease	0.40(0.98) = 0.392	0.008	0.40
	0.950	0.050	

The entries in the Not Correct column were found by subtraction.

$$P(\text{decrease}|\text{not correct}) = \frac{P(\text{decrease} \cap \text{not correct})}{P(\text{not correct})} = \frac{0.008}{0.050} = 0.160$$

R.4   **a.** The mean and standard deviation may be computed using a calculator or computer package. We obtained a mean of 1.794 and a standard deviation of 1.333. If we apply the Empirical Rule, 95% of the values should be in the range $1.794 \pm 2(1.333)$. Note that this range includes negative values, impossible to have for bad debts. The Empirical Rule doesn't work well here.

b. Outliers are indicated in the Minitab stem-and-leaf displays as LO or HI. In this exercise, the only outliers are on the HI side. These show one outlier in Deposits, one in Capital, one in Reserves, and two in BadDebts. It wouldn't make much sense to think of an outlier in the TypeBank variable, because it's qualitative.

R.7   **a.** The output shows that $\bar{x}_3 = 2.31$ and $s = 1.38$, rounded off.

b. The mean and standard deviation for this variable are really meaningless. The numbers shown are arbitrary, qualitative codes. The "average type of meat" just doesn't make sense.

R.8   We must calculate differences $Y = X_5 - X_4$ for all observations. The differences are

5,000	12,700	11,300	1,300	4,000	14,300	15,900
1,200	14,800	1,300	16,700	3,900	2,100	

By standard calculations, we get

Variable	X_4	X_5	Y
Mean	32,938	40,977	8,038
St. dev.	21,561	26,433	6,256
Median	21,200	36,000	5,000

Except for roundoff error, $\bar{y} = \bar{x}_5 - \bar{x}_4$. The median of the differences is not the difference of the medians, nor is the standard deviation of the differences equal to the difference of the standard deviations.

R.9   One measure of skewness is

$$\frac{\bar{y}-\text{median}}{s} = \frac{8,038-5,000}{6,256} = 0.486$$

This large positive value indicates considerable right skewness.
    To obtain a visual impression of the data, we obviously need a plot of the data. A histogram is shown here

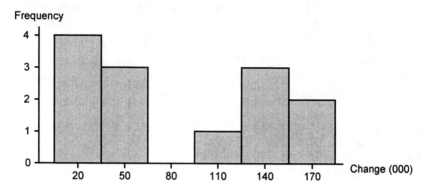

There is some evidence of right skewness, though not as severe skewness as the measure indicates.

R.10   We are trying to calculate P(bottom 1/3 and not canceled). Because this is an "and" probability, we should try the multiplication principle. There is more information than we really need in the exercise. We have P(bottom 1/3) = 0.50 and P(canceled|bottom 1/3) = 0.85. We can find P(not canceled|bottom 1/3) by complements.

P(bottom 1/3 and not canceled) = P(bottom 1/3)P(not canceled|bottom 1/3)
$$= 0.50(1-0.85) = 0.075$$

R.11   No, independence was not assumed. We found that P(bottom 1/3|Yes) = 0.724 but P(bottom 1/3) = 0.50. Independence would mean that the chances of cancellation are the same for high-rated shows as for low-rated shows. Of course, that's not true.

R.12   **a.** We are trying to find P(excellent $\cap$ definite), so we want to use the multiplication principle. The information we have includes only unconditional probabilities. With only the given information, we must calculate

P(excellent $\cap$ definite) = P(excellent)P(definite)
$$= (0.18)(0.24) = 0.0432$$

21

b. Because we had very limited information, we had to assume that performance and potential were independent. Instead of the unknown conditional probability $P(\text{definite}|\text{excellent})$ that we should have used, we had to use the unconditional probability $P(\text{definite})$. The assumption of independence in this situation seems highly unreasonable. Excellent performers should have a higher probability of a definite potential rating. The calculated probability is very likely to be too low.

R.19   **a.** In each case, there is slight right skewness. In the boxplots, the bottom "whisker" is shorter than the top one, and the bottom half of the box is narrower than the top half. The combined data in the histogram is bimodal. The much longer times are from the wheat cleanout. Notice that the boxplots don't overlap at all.

b. The means are 11.250, 12.972, and 32.138, respectively. The medians are 10.75, 12.25, and 31, respectively. In each case the mean is somewhat larger than the median, indicating a degree of right skewness.

R.21   **a.** In looking for a trend, we want to see if there is a noticeable drift upward or downward as we move from left to right in the chart. There isn't anything obvious in the means chart. Perhaps the means at the right are very slightly lower. We don't see any obvious trend.

b. We certainly don't see any trend at all in the S chart. The standard deviations stay very close to constant.

c. With the exception of one value, the process seems very stable and in control.

R.22   This is a special cause, a "one-shot" variation that has an identifiable, unusual cause. In fact, the variation is mostly due to the definition, rather than any change in performance. Rather than resulting from a larger number of patients, the change in mean was the result of a decrease in the base.

Chapter 4

Random Variables and Probability Distributions

4.1 Random Variable: Basic Ideas
4.2 Probability Distributions: Discrete Random Variables

4.6   **a.** The height of each rectangle in a probability histogram should be in proportion to its probability. A probability histogram for Y is shown below:

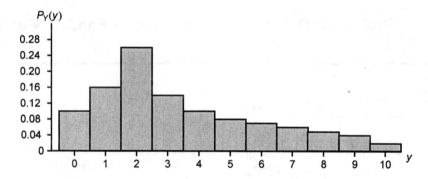

Note the right skewness (long tail toward the higher values).

b. We add probabilities for all relevant cases.

$$P(Y \leq 2) = P(Y = 2) + P(Y = 1) + P(Y = 0)$$
$$= 0.25 + 0.15 + 0.10 = 0.50$$

c. Again, we simply add probabilities.

$$P(Y \geq 7) = P(Y = 7) + P(Y = 8) + P(Y = 9) + P(Y = 10)$$
$$= 0.050 + 0.040 + 0.025 + 0.015 = 0.130$$

d. Again, add probabilities for all cases.

$$P(1 \le Y \le 5) = P(Y = 1) + P(Y = 2) + P(Y = 3) + P(Y = 4) + P(Y = 5)$$
$$= 0.15 + 0.25 + 0.14 + 0.09 + 0.08 = 0.71$$

4.7 The cumulative probabilities are found by adding individual probabilities, starting with the smallest possible value, in this case $y = 0$. Refer to Exercise **4.6** for the individual probabilities. For example, the cdf for $y = 2$ is $0.100 + 0.150 + 0.200 = 0.500$. The cdf of Y is shown below:

y	0	1	2	3	4	5	6	7	8	9	10
$F_Y(y)$	0.10	0.25	0.50	0.64	0.73	0.81	0.87	0.92	0.96	0.985	1.00

$$P(Y \le 2) = F_Y(2) = 0.50$$
$$P(Y \ge 7) = 1 - F_Y(6) = 1 - 0.87 = 0.13$$
$$P(1 \le Y \le 5) = F_Y(5) - F_Y(0) = 0.81 - 0.10 = 0.71$$

4.3   Probability Distributions: Continuous Random Variables

$(\partial, \int)$

4.13   **a.**

Y	F(y)
0.00	0.0000
0.25	0.2801
0.50	0.4630
0.75	0.5625
1.00	0.5926
1.25	0.5671
1.50	0.5000
1.75	0.4051
2.00	0.2963
2.25	0.1875
2.50	0.0926
2.75	0.0255
3.00	0.0000

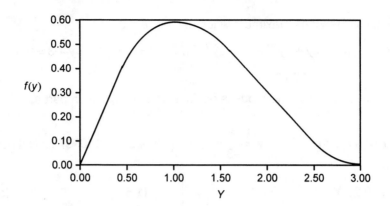

b.
$$P(Y \le 1.5) = \int_{0.0}^{1.5} \frac{4}{27}\left(9y - 6y^2 + y^3\right)dy = \left(\frac{2}{3}y^2 - \frac{8}{27}y^3 + \frac{1}{27}y^4\right)\Big|_{0.0}^{1.5}$$
$$= (1.5 - 1 + 0.1875) - 0 = 0.6875$$

$$P(Y \ge 2.0) = \int_{2.0}^{3.0} \frac{4}{27}\left(9y - 6y^2 + y^3\right)dy = \left(\frac{2}{3}y^2 - \frac{8}{27}y^3 + \frac{1}{27}y^4\right)\Big|_{2.0}^{3.0}$$
$$= 1 - 0.8889 = 0.1111$$

$$P(1 \le Y \le 2.5) = \int_{1.0}^{2.5} \frac{4}{27}\left(9y - 6y^2 + y^3\right)dy = \left(\frac{2}{3}y^2 - \frac{8}{27}y^3 + \frac{1}{27}y^4\right)\Big|_{1.0}^{2.5}$$
$$= 0.9838 - 0.7037 = 0.2801$$

c. $\quad F(y) = \int_{0.0}^{y} \frac{4}{27}\left(9t - 6t^2 + t^3\right)dt = \frac{2}{3}\left(t^2 - \frac{8}{27}t^3 + \frac{1}{27}t^4\right)\Big|_{0}^{y} = \frac{1}{27}\left(18y^2 - 8y^3 + y^4\right)$

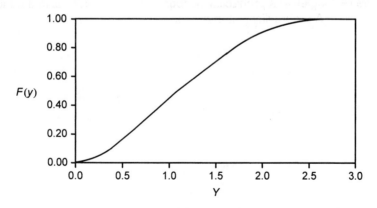

4.14 The density reaches maximum when derivative = 0.

$$f'(y) = \frac{4}{27}\left(9 - 12y + 3y^2\right) = 0 \Rightarrow 3y^2 - 12y + 9 = 0 \Rightarrow y^2 - 4y + 3 = 0$$
$$\Rightarrow (y-1)(y-3) = 0$$

$f(1) = 0.59$, $f(3) = 0$. The mode is obvious from the graph in part **a**.

4.20 **a.** $P(2 \le Y \le 4) = \int_{2}^{4} 3y^{-4}dy = \frac{3}{-3}y^{-3}\Big|_{2}^{4} = -\left(4^{-3}\right) + 2^{-3} = 0.109375$. Not at all important.

b. $P(0.5 \le Y \le 1.5) = \int_{1}^{1.5} 3y^{-4}dy = \frac{3}{-3}y^{-3}\Big|_{1}^{1.5} = -\left(1.5^{-3}\right) + 1^{-3} = 0.7037$. Note that density is only defined on $1 < y < \infty$.

4.21 **a.** $F(y) = \int_{1}^{y} 3t^{-4}dt = -\left(t^{-3}\right)\Big|_{1}^{y} = -y^{-3} + 1^{-3} = 1 - y^{-3}$, $1 < y < \infty$

b. $0.99 = 1 - y^{-3} \Rightarrow y^{-3} = 0.01 \Rightarrow y^3 = 100 \Rightarrow y = \sqrt[3]{100} = 4.642$

4.4 Expected Value and Standard Deviation: Discrete Random Variables

4.22 **a.** The rectangles in a probability histogram are proportional to the individual probabilities. A probability histogram for Y is shown below:

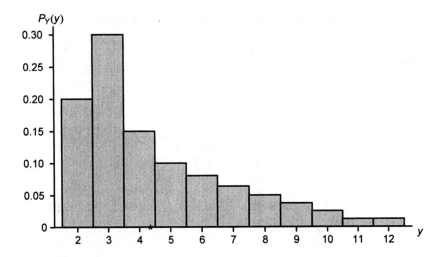

b. The expected value of Y can be calculated as a probability weighted average. Multiply each possible value by its probability and sum.

$$E(Y) = 2(0.2) + 3(0.3) + 4(0.15) + 5(0.1) + 6(0.8) + 7(0.6) + 8(0.04)$$
$$+9(0.03) + 10(0.02) + 11(0.01) + 12(0.01)$$
$$= 0.4 + 0.9 + 0.6 + 0.5 + 0.48 + 0.42 + 0.32 + 0.27 + 0.2 + 0.11 + 0.12 = 4.32$$

c. A star (*) marks $E(Y) = 4.32$ on the histogram. $E(Y)$ is pulled to the right, away from the bulk of the data, because of the right-skewness of the histogram.

4.23　　**a.** To calculate the standard deviation of Y, use the definition,

$$\sigma_Y = \sqrt{\operatorname{Var}(Y)}$$

where

$$\operatorname{Var}(Y) = \sum_{\text{all } y} (y - \mu_Y)^2 P_Y(y) \qquad \mu_Y = E(Y)$$

The following columns of computations facilitate the computation:

y	$P_Y(y)$	$(y - \mu_Y)$	$(y - \mu_Y)^2$	$(y - \mu_Y)^2 P_Y(y)$
2	0.20	−2.32	5.3824	1.076480
3	0.30	−1.32	1.7424	0.522720
4	0.15	−0.32	0.1024	0.015360
5	0.10	0.68	0.4624	0.046240
6	0.08	1.68	2.8224	0.225792
7	0.06	2.68	7.1824	0.430944
8	0.04	3.68	13.5424	0.541696
9	0.03	4.68	21.9024	0.657072
10	0.02	5.68	32.2624	0.645248
11	0.01	6.68	44.6224	0.446224
12	0.01	7.68	58.9824	0.589824
			Total	5.197600

Therefore,

$$Var(Y) = 5.19758$$

so

$$\sigma_Y = \sqrt{5.19758} = 2.2798$$

b. Using the shortcut method

$$Var(Y) = \sum_{all\ y} y^2 P_Y(y) - \mu_Y^2$$

The following columns of computation facilitate the computation:

y	$P_Y(y)$	y^2	$y^2 P_Y(y)$
2	0.20	4	0.80
3	0.30	9	2.70
4	0.15	16	2.40
5	0.10	25	2.50
6	0.08	36	2.88
7	0.06	49	2.94
8	0.04	64	2.56
9	0.03	81	2.43
10	0.02	100	2.00
11	0.01	121	1.21
12	0.01	144	1.44
		Total	23.86

28

Therefore

$$\text{Var}(Y) = 23.86 - (4.32)^2 = 5.1976$$

so that

$$\sigma_Y = \sqrt{5.19758} = 2.2798$$

4.24 Refer to Exercises **4.22** and **4.23**. Recall that $E(Y) = 4.32$ and $\sigma_Y = 2.2798$. The interval within $1\sigma_Y$ of $E(Y)$ is

$$E(Y) \pm 1\sigma_Y = 4.32 \pm 2.2798$$

or

$$2.0402 \text{ to } 6.5998$$

The actual probability is

$$P(2.0402 \le Y \le 6.5998) = P(Y = 3) + P(Y = 4) + P(Y = 5) + P(Y = 6)$$
$$= 0.30 + 0.15 + 0.10 + 0.08 = 0.63$$

The probability based on the Empirical Rule is

$$P(2.0402 \le Y \le 6.5998) = 0.68$$

The discrepancy between the actual probability and the Empirical Rule estimate is due in part to the right-skewness of the population distribution, $P_Y(y)$. The Empirical Rule holds for mound-shaped probability histograms. Also, the probability histogram is quite discrete; the Empirical Rule works best when there are many possible values.

4.27 **a.** We want μ_x, the expected value of X. It is the probability-weighted average of the possible values.

$$\mu_x = \sum x P_x(X) = 0(0.06) + 1(0.14) + \cdots + 10(0.03) = 3.97$$

b. By definition, $\text{Var}(X)$ is the sum of squared deviations from the mean, each multiplied by its probability.

$$\sigma_x^2 = \sum (x - \mu_x)^2 P_x(x) = (0 - 3.97)^2(0.06) + (1 - 3.97)^2(0.14) + \cdots + (10 - 3.97)^2(0.03) = 7.0891$$

c. The shortcut method

$$\sigma_x^2 = \left[\sum x^2 P_x(x)\right] - \mu_x^2$$

requires that we square each value, multiply by the probability, and sum; at the end, subtract the square of the mean.

$$\sum x^2 P_x(x) = 0^2(0.06) + 1^2(0.14) + \cdots + 10^2(0.03) = 22.85$$

Then

$$\sigma_x^2 = 22.85 - (3.97)^2 = 7.0981$$

4.28 To use the Empirical Rule, we need the previously calculated standard deviation and mean.

$$\sigma_x = \sqrt{\sigma_x^2} = \sqrt{7.0981} = 2.66 \text{ and } \mu_x = 3.97$$

$P(X \text{ is within 2 standard deviations of its mean})$
$= P(\mu_x - 2\sigma_x \le X \le \mu_x + 2\sigma_x) = P(3.97 - 2(2.66) \le X \le 3.97 + 2(2.66))$
$= P(-1.35 \le X \le 9.29)$

The possible x values between −1.35 and 9.29 are $x = 0, 1, \ldots, 9$. (Of course, in this case negative values of x aren't possible.) By adding the corresponding $P_x(x)$ values we find

$$P(-1.35 \le X \le 9.29) = 0.97$$

This probability is fairly close to the Empirical Rule approximation, 0.95.

4.37 **a.** Because the firm holds 30% of the market, we assume that the probability of winning any one bid is 0.30. The company can win $y = 3, 2, 1,$ or 0 bids. To have Y come out equal to 3, the company must win all the bids.

$$P(Y = 3) = P(WWW) = (0.30)(0.30)(0.30) = 0.027$$

Similarly,

$$P(Y = 0) = P(LLL) = (0.70)(0.70)(0.70) = 0.343$$

The event $Y = 1$ can occur in three different ways, corresponding to the choice of which of the three bids wins.

$P(Y = 1) = P(WLL) + P(LWL) + P(LLW)$
$= (0.30)(0.70)(0.70) + (0.70)(0.30)(0.70) + (0.70)(0.70)(0.30) = 0.441$

Similarly,

$$P(Y = 2) = P(WWL) + P(WLW) + P(LWW)$$
$$= (0.30)(0.30)(0.70) + (0.30)(0.70)(0.30) + (0.70)(0.30)(0.30) = 0.189$$

b. We assumed that the probability of winning was the same 0.30 on all three bids and that whether or not one bid won didn't change the probability that any other bid would win. If there are no systematic differences among the bids, we can't think of any reason why the probability would change. Barring collusion, there's no obvious reason for dependence, either. The assumptions seem at least fairly reasonable.

4.38 $E(Y) = 0P(Y = 0) + 1P(Y = 1) + 2P(Y = 2) + 3P(Y = 3)$
$$= 0(0.343) + 1(0.441) + 2(0.189) + 3(0.027) = 0.900$$
By the short-cut method,

$$Var(Y) = 0^2(0.343) + 1^2(0.441) + 2^2(0.189) + 3^2(0.027) - (0.900)^2 = 0.63$$

4.39   The expected value shouldn't change. On average, the company should still win 30% of all its bids, and 30% of 3 is 0.9, as we found in the previous exercise. What should change is the variance. The statement of the exercise indicates that there's a higher probability that the company will either win all its bids or lose all its bids, that is, that Y will come out either 3 or 0. Therefore, there must be a smaller probability that Y will come out 1 or 2. If there is higher probability on the more extreme values, the variance must be larger.

4.5   Expected Value and Standard Deviation: Continuous Random Variables $\left(\int \right)$

4.40   **a.** $E(y) = \int_1^\infty y \cdot 3y^{-4}dy = \left. \frac{-3}{2} y^{-2} \right|_1^\infty = 0 + \frac{3}{2} = 1.5$. The long-run average number of worker hours required to assemble an item is 1.5 hours.

b. $E(y^2) = \int_1^\infty y^2 \cdot 3y^{-4}dy = \left. \frac{-3}{1} y^{-1} \right|_1^\infty = 0 + 3 = 3$.

$Var(y) = E(y^2) - E(y)^2 = 3 - (1.5)^2 = 0.75$. So, standard deviation $= 0.866$.

4.41   **a.**

Y	$f(y)$
1.0	3.0000
1.5	0.5926
2.0	0.1875
2.5	0.0768
3.0	0.0370

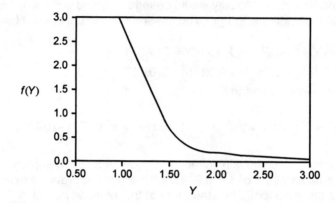

b. Empirical Rule shouldn't work well as distribution is not symmetric and bell-shaped. Within 1 standard deviation of the mean $= 1.5 \pm 0.866 = (0.634, 2.366)$.

P(within 1 standard deviation of the mean) $= P(0.634 \le Y \le 2.366) = P(1 \le Y \le 2.366)$

$$= \int_{1}^{2.366} 3y^{-4}dy = -y^{-3}\Big|_{1}^{2.366} = -0.0755 + 1 = 0.9245.$$

Much larger than 68%, so rule doesn't work well here.

4.45   **a.** 9.399956

b. $\text{Var}(X) = E(X^2) - E(X)^2 = 100.00 - (9.399856)^2 = 11.6427$. So $\sigma_X = 3.412$.

4.6   Joint Probability Distributions and Independence

4.46   **a.** To find $P(X = 1, Y = 2)$, use the entry in the first row and second column of the given joint probability table. By definition

$$P(X = 1, Y = 2) = P_{XY}(1, 2) = 0.055 .$$

b. There are four cases with $X \leq 2, Y \leq 2$.

$$P(X \leq 2, Y \leq 2) = P(X = 1, Y = 1) + P(X = 1, Y = 2) + P(X = 2, Y = 1) + P(X = 2, Y = 2)$$

Using the joint probability table, identify and sum the above 4 joint probabilities, so that

$$P(X \leq 2, Y \leq 2) = 0.030 + 0.055 + 0.055 + 0.070 = 0.21$$

c. To find the probability $P_X(x)$, add up the joint probabilities of that x value and each possible y value.

$$P_X(x) = \sum_{\text{all } y} P_{XY}(x, y)$$

$P_X(x)$ for each value of x is given below:

$$P(X = 1) = \sum_{y=1}^{4} P_{XY}(1, y) = P_{XY}(1, 1) + P_{XY}(1, 2) + P_{XY}(1, 3) + P_{XY}(1, 4)$$

$$= 0.030 + 0.055 + 0.070 + 0.075 = 0.23$$

$$P(X = 2) = \sum_{y=1}^{4} P_{XY}(2, y) = P_{XY}(2, 1) + P_{XY}(2, 2) + P_{XY}(2, 3) + P_{XY}(2, 4)$$

$$= 0.055 + 0.070 + 0.075 + 0.070 = 0.27$$

$$P(X = 3) = \sum_{y=1}^{4} P_{XY}(3, y) = P_{XY}(3, 1) + P_{XY}(3, 2) + P_{XY}(3, 3) + P_{XY}(3, 4)$$

$$= 0.070 + 0.075 + 0.070 + 0.055 = 0.27$$

$$P(X = 4) = \sum_{y=1}^{4} P_{XY}(4, y) = P_{XY}(4, 1) + P_{XY}(4, 2) + P_{XY}(4, 3) + P_{XY}(4, 4)$$

$$= 0.075 + 0.070 + 0.055 + 0.030 = 0.23$$

In the same way, the marginal probabilities, $P_Y(y)$, for y can be computed as

$$P_Y(y) = \sum_{\text{all } y} P_{XY}(x, y)$$

$P_Y(y)$ for each value of y is given below:

$$P(Y = 1) = \sum_{x=1}^{4} P_{XY}(x, 1) = P_{XY}(1, 1) + P_{XY}(2, 1) + P_{XY}(3, 1) + P_{XY}(4, 1)$$

$$= 0.030 + 0.055 + 0.070 + 0.075 = 0.23$$

$$P(Y = 2) = \sum_{x=1}^{4} P_{XY}(x, 2) = P_{XY}(1, 2) + P_{XY}(2, 2) + P_{XY}(3, 2) + P_{XY}(4, 2)$$

$$= 0.055 + 0.070 + 0.075 + 0.070 = 0.27$$

$$P(Y = 3) = \sum_{x=1}^{4} P_{XY}(x, 3) = P_{XY}(1, 3) + P_{XY}(2, 3) + P_{XY}(3, 3) + P_{XY}(4, 3)$$

$$= 0.070 + 0.075 + 0.070 + 0.055 = 0.27$$

$$P(Y = 4) = \sum_{x=1}^{4} P_{XY}(x, 4) = P_{XY}(1, 4) + P_{XY}(2, 4) + P_{XY}(3, 4) + P_{XY}(4, 4)$$

$$= 0.075 + 0.070 + 0.055 + 0.030 = 0.23$$

d. X and Y are statistically independent if (and only if)

$$P_{XY}(x, y) = P_X(x)P_Y(y) \quad \text{for } all \ x, y$$

For the given joint probability table, check the first cell probability, $P_{XY}(1, 1)$:

$$P_{XY}(1, 1) = 0.030 \neq P_X(1)P_Y(1) = 0.23(0.23) = 0.0529$$

Therefore, X and Y are *not* statistically independent.

4.47   To show that

$$P_{XY}(x, y) = 0.005(-10 + 10x + 10y - x^2 - y^2 - 2xy)$$

yields the joint probability table of Exercise **4.66**, simply evaluate this expression for the appropriate x and y values as shown below:

$$P_{XY}(1, 1) = 0.005\left[-10 + 10(1) + 10(1) - 1^2 - 1^2 - 2(1)(1)\right] = 0.030$$
$$P_{XY}(1, 2) = 0.005\left[-10 + 10(1) + 10(2) - 1^2 - 2^2 - 2(1)(2)\right] = 0.055$$
$$P_{XY}(1, 3) = 0.005\left[-10 + 10(1) + 10(3) - 1^2 - 3^2 - 2(1)(3)\right] = 0.070$$
$$P_{XY}(1, 4) = 0.005\left[-10 + 10(1) + 10(4) - 1^2 - 4^2 - 2(1)(4)\right] = 0.075$$
$$P_{XY}(2, 1) = 0.005\left[-10 + 10(2) + 10(1) - 2^2 - 1^2 - 2(2)(1)\right] = 0.055$$
$$P_{XY}(2, 2) = 0.005\left[-10 + 10(2) + 10(2) - 2^2 - 2^2 - 2(2)(2)\right] = 0.070$$
$$P_{XY}(2, 3) = 0.005\left[-10 + 10(2) + 10(3) - 2^2 - 3^2 - 2(2)(3)\right] = 0.075$$
$$P_{XY}(2, 4) = 0.005\left[-10 + 10(2) + 10(4) - 2^2 - 4^2 - 2(2)(4)\right] = 0.070$$
$$P_{XY}(3, 1) = 0.005\left[-10 + 10(3) + 10(1) - 3^2 - 1^2 - 2(3)(1)\right] = 0.070$$
$$P_{XY}(3, 2) = 0.005\left[-10 + 10(3) + 10(2) - 3^2 - 2^2 - 2(3)(2)\right] = 0.075$$
$$P_{XY}(3, 3) = 0.005\left[-10 + 10(3) + 10(3) - 3^2 - 3^2 - 2(3)(3)\right] = 0.070$$
$$P_{XY}(3, 4) = 0.005\left[-10 + 10(3) + 10(4) - 3^2 - 4^2 - 2(3)(4)\right] = 0.055$$
$$P_{XY}(4, 1) = 0.005\left[-10 + 10(4) + 10(1) - 4^2 - 1^2 - 2(4)(1)\right] = 0.075$$
$$P_{XY}(4, 2) = 0.005\left[-10 + 10(4) + 10(2) - 4^2 - 2^2 - 2(4)(2)\right] = 0.070$$
$$P_{XY}(4, 3) = 0.005\left[-10 + 10(4) + 10(3) - 4^2 - 3^2 - 2(4)(3)\right] = 0.055$$
$$P_{XY}(4, 4) = 0.005\left[-10 + 10(4) + 10(4) - 4^2 - 4^2 - 2(4)(4)\right] = 0.030$$

To find an expression for $P_X(x)$, use the basic relation

$$P_X(x) = \sum_{\text{all } y} P_{XY}(x, y)$$

so that,

$$P_X(x) = \sum_{y=1}^{4} 0.005\left(-10 + 10x - 10y - x^2 - y^2 - 2xy\right)$$

$$= 0.005\left[-10 + 10x + 10(1) - x^2 - (1)^2 - 2x(1)\right] + 0.005\left[-10 + 10x + 10(2) - x^2 - (2)^2 - 2x(2)\right]$$

$$+0.005\left[-10 + 10x + 10(3) - x^2 - (3)^2 - 2x(3)\right] + 0.005\left[-10 + 10x + 10(4) - x^2 - (4)^2 - 2x(4)\right]$$

$$= 0.005\left(-40 + 40x + 100 - 4x^2 - 30 - 20x\right) = 0.005\left(30 + 20x - 4x^2\right)$$

4.7   Covariance and Correlation of Random Variables

4.56   **a.** The marginal probability distributions of X and Y are both symmetric around the value 2. Therefore $\mu_X = \mu_Y = 2$.

b. Using the shortcut method for calculating variances, we square values, weight by probabilities and sum, subtracting the square of the mean at the end.

$$\sigma_X^2 = 0^2(0.180) + 1^2(0.190) + \cdots + 4^2(0.180) - 2^2 = 1.83$$

$$\sigma_X = \sqrt{1.83} = 1.353$$

Similarly

$$\sigma_Y^2 = 1.65 \text{ and } \sigma_Y = \sqrt{1.65} = 1.285$$

4.57   **a.** The shorter way to find Cov(X, Y) is

$$\sum_X \sum_Y xy P_{XY}(x, y) - \mu_X \mu_Y$$

We have $\mu_X = \mu_Y = 2$

$$\text{Cov}(X, Y) = 0 \cdot 0 \cdot (0.010) + 0 \cdot 1 \cdot (0.015) + \cdots$$

$$+ \cdots$$

$$+ \cdots 4 \cdot 3 \cdot (0.015) + 4 \cdot 4 \cdot (0.010) - 2 \cdot 2$$

$$= 3.30 - 4 = -0.70$$

b. From Exercise **4.56**, $\sigma_X = 1.353$ and $\sigma_Y = 1.285$

$$\text{Corr}(X, Y) = \rho_{XY} = \frac{\text{Cov}(X, Y)}{\sigma_X \sigma_Y} = \frac{0.70}{(1.353)(1.285)} = -0.403$$

As X increases, Y tends to decrease, as shown by the negative correlation. X and Y can't be independent, because independent random variables have 0 correlation.

4.58   The conditional expectation of Y given $X = x$ is calculated like any expected value, but using conditional probabilities. For example, the conditional probabilities given were found in Exercise **4.52** to be 0.0556, 0.0833, 0.1667, 0.4167, and 0.2778 for $y = 0, 1, 2, 3,$ and 4 respectively. Thus

$$\mu_{Y|X=x} = 0(0.0556) + 1(0.0833) + \cdots + 4(0.2778) = 2.778$$

Similar calculations lead to the following table:

x	0	1	2	3	4
$\mu_{Y|X=x}$	2.778	2.359	2.000	1.641	1.222

As x increases, $\mu_{Y|X=x}$ decreases, as we'd expect given that X and Y have a negative correlation.

4.59   **a.** The addition law can be used to find $P_T(t)$. For example $T = X + Y = 2$ if ($X = 0$ and $Y = 2$) or ($X = 1$ and $Y = 1$) or ($X = 2$ and $Y = 0$). Thus

$$P_T(2) = P_{XY}(0, 2) + P_{XY}(1, 1) + P_{XY}(2, 0) = 0.030 + 0.030 + 0.030 = 0.090$$

Similar calculations yield the distribution of T

t	0	1	2	3	4	5	6	7	8
$P_T(t)$	0.010	0.035	0.090	0.205	0.320	0.205	0.090	0.035	0.010

b. $\mu_T = 4$, by symmetry.
To calculate the variance, the shortcut approach is convenient.

$$\sigma_T^2 = 0^2(0.010) + 1^2(0.035) + \cdots + 8^2(0.010) - 4^2 = 18.08 - 16 = 2.08$$

The variance could equally well be calculated from the definition.

c. We previously have found all the relevant means, variances, and the covariance. We must have $\mu_T = \mu_X + \mu_Y = 2 + 2 = 4$ and

$$\sigma_T^2 = \sigma_X^2 + \sigma_Y^2 + 2\text{Cov}(X, Y) = 1.83 + 1.65 + 2(-0.70) = 2.08$$

4.62   **a.** Expected return $= 10000(0.085) + 10000(0.08) = 850 + 800 = 1650$.
Variance $= 10000^2(0.005) + 10000^2(0.0045) + 2(10000)(10000)(0.0035) = 1650000$.

b. For 1 and 4: Variance
$= 10000^2(0.005) + 10000^2(0.0032) + 2(10000)(10000)(0.0031) = 1440000$.
For 1 and 3: Variance
$= 10000^2(0.005) + 10000^2(0.004) + 2(10000)(10000)(0.0006) = 1020000$.
1 and 3 have the smaller risk (variance).

4.8   Joint Probability Densities for Continuous Random Variables $\left(\int \right)$

4.67   **a.**

y	f(y)
0	0
0.1	0.0027
0.2	0.0384
0.3	0.1701
0.4	0.4608
0.5	0.9375
0.6	1.5552
0.7	2.1609
0.8	2.4576
0.9	1.9683
1.0	0

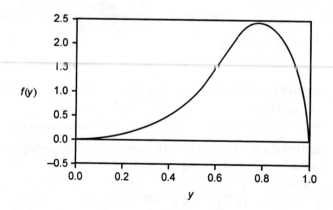

b. $P(0.7 < Y < 0.9) = \int_{0.7}^{0.9} 30y^4(1-y)dy = 30\left[\dfrac{y^5}{5} - \dfrac{y^6}{6}\right]_{0.7}^{0.9} = 0.46556$

c. $P(Y > 0.8) = \int_{0.8}^{1.0} 30y^4(1-y)dy = 30\left[\dfrac{y^5}{5} - \dfrac{y^6}{6}\right]_{0.8}^{1.0} = 1 - 0.65536 = 0.34464$

4.68 $E(Y) = \int_{0.0}^{1.0} y \cdot 30y^4(1-y)dy = 30\left[\dfrac{y^6}{6} - \dfrac{y^7}{7}\right]_{0.0}^{1.0} = 0.714286$

$E(Y^2) = \int_{0.0}^{1.0} y^2 \cdot 30y^4(1-y)dy = 30\left[\dfrac{y^7}{7} - \dfrac{y^8}{8}\right]_{0.0}^{1.0} = 0.535714$

$Var(Y) = E(Y^2) - E(Y)^2 = 0.535714 - (0.714286)^2 = 0.02551$. So, $\sigma_Y = 0.159718$.

4.71 **a.** No, X and Y are not independent.

b. The correlation is 0.00084301, so the strength of the relationship is very weak.

Supplementary Exercises

4.75 **a.** One good way to find the probability distribution is to create a tree. Begin branching on whether or not automatic transmission is ordered, because we have unconditional probabilities for that. Then branch on air conditioning or not, and finally on interior decor or not. Label each path of the tree according to how many packages are ordered along that path.

Automatic transmission?	Air conditioning?	Interior decor?		y probability	
		Y	0.40	3	0.192
	Y 0.60	N	0.60	2	0.288
Y 0.80		Y	0.30	2	0.096
	N 0.40	N	0.70	1	0.224
		Y	0.30	2	0.030
	Y 0.50	N	0.70	1	0.070
N 0.20		Y	0.20	1	0.020
	N 0.50	N	0.80	0	0.080

Adding up probabilities for various y values, we obtain the probability distribution

y	0	1	2	3
$P_Y(y)$	0.080	0.314	0.414	0.192

b. Once again, add probabilities for all relevant cases.

$$P(Y \geq 2) = P(Y = 2) + P(Y = 3) = 0.414 + 0.192 = 0.606$$

c. The cumulative probability distribution $F_Y(y)$ is found by adding probabilities from the lowest possible value, in this case $y = 0$, up to the specified y value. For $y = 0$, the only value to be added is the probability of exactly 0, namely 0.080. For $y = 1$, add the probabilities of $y = 0$ and 1; $F_Y(1) = 0.080 + 0.314 = 0.394$. We obtain the following table of cumulative probabilities.

y	0	1	2	3
$F_Y(y)$	0.080	0.394	0.808	1.000

To find the probability that Y is at least two, use the complements law. The complementary event is $Y \leq 1$. The probability that $Y \leq 1$ is $F_Y(1) = 0.394$. Therefore

$$P(Y \geq 2) = 1 - 0.394 = 0.606$$

once again.

4.76 We can use the definition to find the mean

$$\mu_Y = 0(0.080) + 1(0.314) + 2(0.414) + 3(0.192) = 1.718$$

The shortcut method is a convenient way to find the variance and standard deviation of Y.

$$\sigma_Y^2 = 0^2(0.080) + 1^2(0.314) + 2^2(0.414) + 3^2(0.192) - (1.718)^2 = 0.7465$$
$$\sigma_Y = \sqrt{0.7465} = 0.8640$$

4.77 **a.** A convenient way to attack the problem is to refer to the probability tree in Exercise **4.75**. The dollar profits associated with each path are (from top to bottom of the tree) 450, 350, 300, 200, 250, 150, 100, and 0. All we need to do to find the probability distribution is to copy the probabilities.

x	0	100	150	200	250	300	350	450
$P_X(x)$	0.080	0.020	0.070	0.224	0.030	0.096	0.288	0.192

b. The mean is the probability-weighted average of possible values.

$$\mu_X = \sum x P_X(x) = 0(0.080) + 100(0.020) + \cdots + 450(0.192) = 280.8$$

A convenient way to find the standard deviation is to use the shortcut method for calculating the variance.

$$\sigma_X^2 = \sum x^2 P_X(x) - \mu_X^2 = 0^2(0.080) + 100^2(0.020) + \cdots + 450^2(0.192) - (280.8)^2 = 16{,}561.36$$

$$\sigma_X = \sqrt{16{,}561.36} = 128.69$$

4.81   **a.**

y	$f(y)$
0.70	0.585
0.75	1.406
0.80	2.875
0.85	4.816
0.90	5.989
0.95	3.962
1.00	0.000

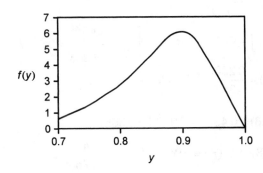

b. $\log f(y) = \log 3990 + 18\log y + 2\log(1-y)$.
$(d\log f(y))/dy = (18/y) - (2/1 - y) = 0 \Rightarrow y = 0.9$. Mode occurs at 0.9. (Recall from calculus that the derivative of $\log y = 1/y$, and the derivative of $\log(1-y) = -1/(1-y)$.)

c. $P(\text{At least 90\% pass}) = \int_{0.09}^{1} 3990 y^{18}(1-y)^2 dy = 3990 \left[\dfrac{y^{19}}{19} - \dfrac{y^{20}}{20} + \dfrac{y^{21}}{21} \right]_{0.9}^{1} = 0.3516$

$P(\text{No more than 85\% pass}) = \int_{0}^{0.85} 3990 y^{18}(1-y)^2 dy = 3990 \left[\dfrac{y^{19}}{19} - \dfrac{y^{20}}{20} + \dfrac{y^{21}}{21} \right]_{0}^{0.85}$

$= 0.3705$

4.82 **a.** $E(Y) = 3990 \int_0^1 y^{19}(1-y)^2 dy = 3990 \left[\dfrac{19!2!}{(19+2+1)!} \right] = 0.8636$

b. $E(Y^2) = 3990 \int_0^1 y^{20}(1-y)^2 dy = 3990 \left[\dfrac{20!2!}{(20+2+1)!} \right] = 0.7510$

$Var(Y) = E(Y^2) - E(Y)^2 = 0.7510 - (0.8636)^2 - 0.005121.$ $\sigma_y = 0.07156$

c. $E(Y) - 3\sigma_y = 0.8636 - 3(0.07156) = 0.649$

$P(Y < 0.649) = \int_0^{0.649} 3990 y^{18}(1-y)^2 dy = 3990 \left[\dfrac{y^{19}}{19} - \dfrac{y^{20}}{20} + \dfrac{y^{21}}{21} \right]_0^{0.649} = 0.008407$

4.88 **a.** $P(X > 0.8) = \int_{0.8}^1 30(x^4 - x^5)dx = 30 \left[\dfrac{x^5}{5} - \dfrac{x^6}{6} \right]_{0.8}^1 = 1 - 0.6554 = 0.3446$

b. $P(X < 0.5) = \int_0^{0.5} 30(x^4 - x^5)dx = 30 \left[\dfrac{x^5}{5} - \dfrac{x^6}{6} \right]_0^{0.5} = 0.1094$

4.89 $P(X < 0.9 | X > 0.8) = \dfrac{P(0.8 < X < 0.9)}{P(X > 0.8)}$

$P(0.8 < X < 0.9) = \int_{0.8}^{0.9} 30(x^4 - x^5)dx = 30 \left[\dfrac{x^5}{5} - \dfrac{x^6}{6} \right]_{0.8}^{0.9} = 0.8857 - 0.6554 = 0.2303$

$P(X < 0.9 | X > 0.8) = \dfrac{0.2303}{0.3446} = 0.6683$

4.97 **a.** $P(0.6 < Y < 0.8) = 0.4362 - 0.0705 = 0.3657$

b. $E(Y) = \dfrac{4}{5} = 0.8.$ $Var(Y) = E(Y^2) - E(Y)^2 = \dfrac{4}{275} = 0.0145$

4.98 $df(y)/dy = 72(7)y^6(1-y) - 72y^7 = 0 \Rightarrow 7(1-y) = y \Rightarrow y = 7/8 = 0.875$. Mean < Mode, so distribution is left skewed.

4.101 **a.** The smallest number of cards she can obtain is $x = 1$. This occurs if the first card she buys is a National Leaguer. It is also possible to have $x = 2$ (an American Leaguer, then a National Leaguer), $x = 3$, $x = 4$, and so on. In principle, there is no absolute maximum number of cards. So $x = 1, 2, 3, 4, ...$

b. $P(X = 1) = P(\text{first card bought is a National Leaguer}) = 0.5$

c. To have $x = 2$, first obtain an American Leaguer (probability 0.5), and then a National Leaguer (probability also 0.5).

d. Multiply the probability of an American Leaguer first times the probability of a National Leaguer second.

$$P(X = 2) = 0.5(0.5) = 0.25$$

In general,

$P(X = x) = P(x - 1 \text{ American Leaguers and then a National Leaguer}) = (0.5)^x$

4.102 On the average, she should buy as many American League cards as National League cards. She buys one NL card, so should also buy one AL card. $E(X)$ should be 2.

4.103 The theoretical probabilities are 0.5, 0.25, 0.125, 0.0625, and so on. The simulation results are 0.501, 0.255, 0.131, 0.53, and so on. In general the results are fairly close. The mean (long run average) should be 2; in the simulation (average over 1,000 trials, not an infinite number), the mean is 1.94, which is fairly close.

Chapter 5

Some Special Probability Distributions

5.1   Counting Possible Outcomes

5.3   Do we need sequences or subsets? There is no need to consider the sequence (order) in which the glasses are chosen. Therefore, the combination rule applies.

$$\binom{r}{k} = \frac{r!}{k!(r-k)!}$$

Four glasses are selected from a set of 8 glasses, so that $r = 8$ and $k = 4$. Therefore, the number of ways in which the tester can select the glasses is

$$\binom{8}{4} = \frac{8!}{4!4!} = 70$$

5.4   Refer to Exercise **5.3**. Again, we now must count in two categories, correct glasses and incorrect ones.

To obtain all samples with exactly three correct glasses, we can combine any of the $\binom{4}{3}$ choices of 3 from the 4 correct (private labelled) glasses with any of the $\binom{4}{1}$ choices of 1 from the 4 incorrect (nationally-advertised) glasses. Therefore, there are $\binom{4}{3}\binom{4}{1} = 16$ ways in which the tester can select 3 correct and 1 incorrect glass.

5.2 Bernoulli Trials and the Binomial Distribution

5.5   **a.** This is a straightforward exercise in using a table (or if desired a computer package). We know to look for $n = 10$ and $\pi = 0.2$.

$$P_Y(3) = \binom{10}{3}(0.2)^3(0.8)^{10-3} = \frac{10!}{3!7!}(0.2)^3(0.8)^7 = 120(0.008)(0.2097) = 0.2013$$

as found in Appendix Table 1

b. Again, simply consult Appendix Table 1 or a computer package that calculates binomial probabilities.

$$P_Y(2) = \binom{4}{2}(0.4)^2(0.6)^{4-2} = 0.3456$$

c. Note that with $\pi = 0.7$, we must enter Appendix Table 1 along the bottom and read the y value from the right side.

$$P_Y(12) = \binom{16}{12}(0.7)^{12}(0.3)^{16-12} = 0.2040$$

5.7   **a.** To find $P(Y \geq 4)$ we must add the probabilities of obtaining exactly 4, exactly 5,..., exactly 20 successes. From Appendix Table 1 with $n = 20$ and $\pi = 0.40$

$$P(Y \geq 4) = P_Y(4) + P_Y(5) + \cdots + P_Y(20) = 0.0350 + 0.0746 + \cdots + 0.0000 = 0.9840$$

Note that no probabilities are shown in Table 1 for $y = 19$ and $y = 20$. These probabilities are 0 to four decimal places. This part of the exercise could also be done using a computer package, as could the other parts.

b. In comparison to part **a.**, we do not include $P_Y(4)$

$$P(Y > 4) = P_Y(5) + \cdots + P_Y(20) = P(Y \geq 4) - P_Y(4) = 0.9840 - 0.0350 = 0.9490$$

c. Again, we must add all relevant probabilities.

$$P(Y \leq 10) = P_Y(0) + P_Y(1) + \cdots + P_Y(10) = 0.0000 + 0.0005 + \cdots + 0.1171 = 0.8723$$

d. We don't include the probability of exactly 16 successes here.

$$P(Y > 16) = P_Y(17) + P_Y(18) + P_Y(19) + P_Y(20) = 0.0000$$

5.8 The probabilities may be found by adding probabilities from Appendix Table 1. We look in the $n = 20$ block and find the $\pi = 0.60$ column. Because $\pi > 0.50$ we find $\pi = 0.60$ along the bottom and read y values along the right-hand edge.

$$P(Y \le 16) = P_Y(0) + P_Y(1) + \cdots + P_Y(16) = 0.0000 + 0.0000 + \cdots + 0.0350 = 0.9840$$

A convenient way to find $P(Y \le 16)$ is by subtraction.

$$P(Y < 16) = P(Y \le 16) - P_Y(16) = 0.9840 - 0.0350 = 0.9490$$

These table entries are exactly the ones read in answering parts **a** and **b** of Exercise **5.7**, so the answers are identical. The only difference is that in Exercise **5.7** we chose to call the event with probability 0.40 a success, whereas in Exercise **5.8** we called the complementary event with probability 0.60 a success.

5.12 **a.** The properties of a binomial experiment are listed in the text. The assumption of independence for the parcel service problem is questionable. One non-delivery indicates possible storms, breakdowns, or other factors that would increase the chances of other non-deliveries.

b. Assuming that binomial probabilities apply, we can use Table 1 in the text (or a computer package) to find the probability that Y, the number of packages delivered on time, is greater than or equal to 85. We must add probabilities for all relevant values. Identify $n = 100$, $\pi = 0.90$ and add the entries for $y = 85$, 86, ..., 100 and get 0.9602. Alternatively, find the complementary probability by adding the values for $y = 84$, 83, ..., and get 0.0399, so the probability of 85 or more packages delivered on time is $1 = 0.0399 = 0.9601$. The discrepancy in the fourth decimal place is caused by rounding in calculating the table.

5.13 Although it would be very time consuming with $n = 100$, we could find the expected value of the number of packages delivered on time by using the definition and the probabilities in Table 1. Instead we will use the fact that for binomial probabilities

$$E(Y) = n\pi$$

with $n = 100$ and $\pi = 0.90$
      Therefore,

$$E(Y) = 100(0.90) = 90$$

To find the standard deviation of the number of packages delivered on time we could use the definition and the probabilities found in Table 1. Again we will use an alternative method. We will use the fact that for binomial probabilities

$$\sigma_Y = \sqrt{n\pi(1-\pi)}$$

so that

$$\sigma_Y = \sqrt{100(0.90)(0.10)} = 3$$

5.16 **a.** The probability of 25 or fewer bad addresses is approximated by the proportion of times that the simulation yields 25 or fewer BADADDR values. The output shows the values, listed as Bin Center, Freq (frequencies), Pct (percentages out of the 1000 trials) and Cum Pct (the cumulative percentage less than or equal to the value). The easiest way to answer the question is to note that the cumulative percentage less than or equal to 25 is 2.3%. Therefore, the approximate probability is 0.023.

b. The expected value (really a simulation approximation to that value) is shown as EXPVALUE = 35.685. With an infinitely long simulation, the result would come out as 200(0.18) = 36.000. Similarly, a simulation approximation to the standard deviation is SD = 5.264. The theoretical value is

$$\sqrt{200(0.18)(1-0.18)} = 5.433.$$

5.17 The difficulty with this exercise is that Appendix Table 1 does not contain entries for $\pi = 0.18$. We could go through very tedious hand calculations. A more convenient alternative is to use a computer package. For example, here is Minitab output for the problem.

```
MTB > CDF 25;
SUBC>   Binomial 200 .18.

Cumulative Distribution Function

Binomial with n = 200 and p = 0.180000

        x       P( X <= x)
     25.00        0.0230

MTB > name k1 'mean' k2 'stdev'
MTB > let k1 = 200*.18
MTB > let k2 = sqrt(200*.18*(1-.18))
MTB > print k1 k2

mean      36.0000
stdev     5.43323
```

The output shows that the probability of 25 or fewer successes is 0.0230. The expected value (mean) is 36.0, and the standard deviation is 5.43323.

5.3   The Hypergeometric Distribution

5.19   **a.** $P_Y(y) = \dfrac{\dbinom{3}{y}\dbinom{4}{3-y}}{\dbinom{7}{3}}$, $y = 0, 1, 2, 3$.

Y	$P_Y(y)$
0	0.1143
1	0.5143
2	0.3429
3	0.0285

b.

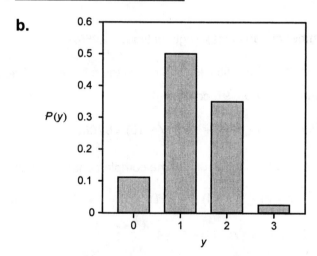

5.20 $E(Y) = (3)(3/7) = 1.2857$

$\mathrm{Var}(Y) = (3)(3/7)(1 - 3/7)((7 - 3)/(7 - 1)) = 0.4898$.   So $\sigma_Y = 0.6999$.

5.23   **a.** $P(\text{selects four correct glasses}) = \dfrac{\dbinom{4}{4}\dbinom{4}{0}}{\dbinom{8}{4}} = 0.01429$.

b.

Y	$P_Y(y)$
0	0.0143
1	0.2286
2	0.5143
3	0.2286
4	0.0143

5.4 Geometric and Negative Binomial Distributions
5.5 The Poisson Distribution

5.29　　Y is the number of claims arising in a 4 week period. Y has a Poisson distribution. We must find the mean for a 4 week period, given a mean of 2.25 for a 1 week period.

$$\mu = \text{(expected number per unit time)(length of time)} = 2.25(4) = 9$$

Table 2, the table of Poisson probabilities, is used to find the following probabilities.

a. Add probabilities found in the $\mu = 9.0$ column.

$$P(Y \leq 10) = P(Y = 0) + P(Y = 1) + \cdots + P(Y = 10) = 0.7060$$

b. We can either add probabilities again or find the complementary probability.

$$P(Y \geq 7) = 1 - P(Y \leq 6) = 1 - [P(Y = 0) + P(Y = 1) + \cdots + P(Y = 6)] = 1 - 0.2067 = 0.7933$$

c.　$P(7 \leq Y \leq 11) = P(Y = 7) + \cdots + P(Y = 11) = 0.5963$

5.30　　The expected value of a Poisson variable is

$$E(Y) = \mu$$

Therefore,

$$E(Y) = 9$$

The standard deviation of a Poisson variable is the square root of the variance, and the variance equals the mean.

$$\sigma_Y = \sqrt{\mu}$$

Therefore,

$$\sigma_Y = \sqrt{9} = 3$$

5.31 There are some none-too-plausible violations. For example, if a fire burns several neighboring houses and two or more of them were insured from this firm, then the assumption that events happen one at a time doesn't hold. Perhaps more likely, fires are often caused by faulty heaters and occur more often in winter; if this holds, the constant rate assumption is wrong.

5.6 The Uniform Distribution

5.38 **a.** $P(300 < Y < 1300) = \dfrac{1000}{9999} = 0.10001$

b. $\text{Var}(Y) = \dfrac{(9999 - 0000)^2}{12} = 8{,}331{,}666.75$

5.7 Exponential Distribution $\left(\int\right)$

5.42 $\lambda = 1.25$ hours/arrival. $P(Y > 1) = 0.2865$, and $P(Y > 2) = 0.0821$ from the exponential distribution.

5.43 Let X = number of arrivals per hour. Then $\mu = 1/1.25 = 0.80$ arrivals/hour.

a. $P(X = 0) = \dfrac{e^{-0.8}(0.8)^0}{0!} = 0.4493$.

b. $\mu = 2(0.80) = 16$ arrivals/2 hours. $P(X = 0) = \dfrac{e^{-1.6}(1.6)^0}{0!} = 0.2019$

c. The exponential distribution fixes the number of arrivals and treats the time between arrivals as random. The Poisson distribution fixes the amount of time and regards the number of arrivals in that time period as random.

5.46 $\lambda = \dfrac{1}{40} = 0.025$

a.

$$P(20 < Y < 60) = \int_{20}^{60} 0.025 e^{-0.025y} dy = e^{-0.025(20)} - e^{-0.025(60)} = 0.6065 - 0.2231 = 0.3834.$$

b. $\text{Var}(y) = \dfrac{1}{\lambda^2}$, so $\sigma_y = \dfrac{1}{\lambda} = \dfrac{1}{0.025} = 40$ days.

5.47 Assumption of constant average rate of occurrence is questionable here.

5.8 The Normal Distribution

5.52 The problems in this exercise are in a sense opposite of the problems in Exercise **5.51**. In that exercise, values were given and probabilities had to be found. Here probabilities are given and values must be found.

a. Below is a picture of the given probability:

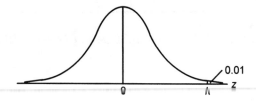

To find k such that $P(Z \geq k) = 0.01$, specify the area $P(0 \leq Z \leq k)$. Because the area to the right of 0 must be 0.5000, $P(0 \leq Z \leq k) = 0.5000 - 0.0100 = 0.4900$. Looking through Table 3 for an area of 0.4900, we find $Z \approx 2.33$. Therefore, $P(Z \geq 2.33) = 0.01$, i.e., $k = 2.33$.

b. Below is a picture of the given probability:

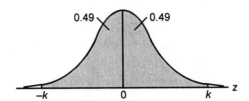

To find k such that $P(-k \leq Z \leq k) = 0.98$, we need only work with the positive half of the curve because of the symmetry property of the normal distribution. An area of 0.4900 has a corresponding Z of 2.33, as we established in part **a.** Therefore, $P(-2.33 \leq Z \leq 2.33) = 0.98$, i.e., $k = 2.33$.

c. Below is a picture of the given probability:

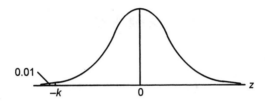

Because the normal curve is symmetric, finding k such that $P(Z \leq -k) = 0.01$ is equivalent to finding $P(Z \geq k) = 0.01$. From part **a.**, we know $P(Z \geq 2.33) = 0.01$, i.e., $k = 2.33$.

d. Below is a picture of the given probability:

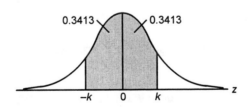

To find k such that $P(-k \leq Z \leq k) = 0.6826$, we need only work with the positive half of the curve because of the symmetry property. So we want to find k such that $P(0 \leq Z \leq k) = 0.6826/2 = 0.3413$. An area of 0.3413 corresponds to $Z = 1$ (from Table 3). Therefore, $P(-1 \leq Z \leq 1) = 0.6826$, i.e., $k = 1$.

e. By symmetry, $P(-k \leq Z \leq 0) = P(0 \leq Z \leq k)$. Each of these probabilities must equal $0.9544/2 = 0.4772$. The entry for $z = 2.00$ is 0.4772, so that $P(-2.00 \leq Z \leq 2.00) = 0.4772 + 0.4772 = 0.9544$. The desired k is 2.00. The required picture is very similar to the one in part **d.**

f. Below is a picture of the given probability:

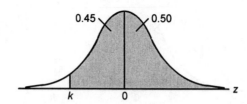

Because of the symmetry of the normal curve, finding k such that $P(Z \geq k) = 0.95$ is equivalent to finding k such that $P(0 \leq Z \leq |k|) = 0.45$. From Table 3, an area of 0.45 corresponds to halfway between $z = 1.64$ and $z = 1.65$. Therefore, $P(Z \geq -1.645) = 0.95$, i.e., $k = -1.645$.

5.53   From Exercise **5.52** parts **d.** and **e.**, we have that

$$P(-1 \leq Z \leq 1) = 0.6826$$

and

$$P(-2 \leq Z \leq 2) = 0.9544$$

From the Empirical Rule, we have that

$$P(-1 \leq Z \leq 1) \approx 0.68$$

and

$$P(-2 \leq Z \leq 2) \approx 0.95$$

The normal curve probabilities are more precise statements of the Empirical Rule probabilities.

5.54 Y is a normally distributed random variable with $\mu = 100$ and $\sigma = 15$.

a. Any normal random variable can be reduced to a standard normal random variable z by subtracting μ and dividing by σ, i.e.,

$$Z = \frac{Y - \mu}{\sigma}$$

Therefore,

$$(Y \leq 130) = \left(\frac{Y - \mu}{\sigma} \leq \frac{130 - 100}{15} \right) = (Z \leq 2)$$

b. $(Y \geq 82.5) = \left(\dfrac{Y-\mu}{\sigma} \geq \dfrac{82.5-100}{15}\right) = (Z \geq -1.17)$

c. To find $P(Y \leq 130)$, first convert the probability statement to its Z score equivalent and then find the probability using Table 3.
  From part **a.**, we know that

$$P(Y \leq 130) = P(Z \leq 2)$$

From Table 3, we find that

$$P(Z \leq 2) = 0.9772$$

Therefore, $P(Y \leq 130) = 0.9772$. See equivalent pictures below:

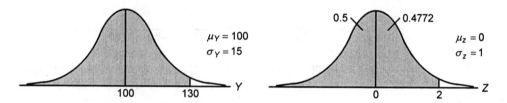

To find $P(Y \geq 82.5)$, note that its Z score equivalent is $P(Z \geq -1.17)$ from part **b.**
  From Table 3, we find that $P(Z \geq -1.17) = 0.8790$

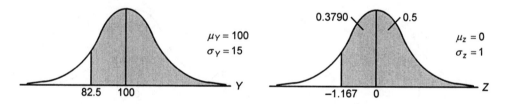

d. For each of the following 3 problems, convert the probability statement to its Z score equivalent and find the probability using Table 3. Pictures accompany each solution.

$$P(Y > 106) = P\left(\dfrac{Y-\mu}{\sigma} > \dfrac{106-100}{15}\right) = P(Z > 0.4) = 0.5 - 0.1554 = 0.3446$$

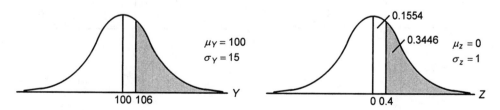

$$P(Y < 94) = P\left(\frac{Y - \mu}{\sigma} < \frac{94 - 100}{15}\right) = P(Z < -0.4) = 0.3446$$

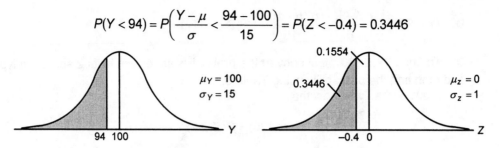

$$P(94 < Y < 106) = P(-0.4 < Z < 0.4) = 0.3108$$

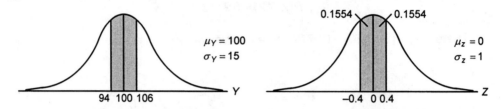

e. For each of the following 3 problems, convert the probability statement to its *Z* score equivalent and find the probability using Table 3. Pictures accompany each solution.

$$P(Y \le 70) = P\left(\frac{Y - \mu}{\sigma} \le \frac{70 - 100}{15}\right) = P(Z \le -2) = 0.0228$$

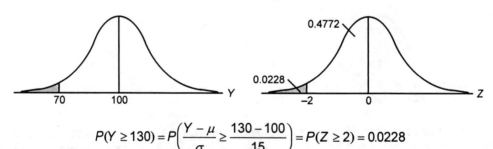

$$P(Y \ge 130) = P\left(\frac{Y - \mu}{\sigma} \ge \frac{130 - 100}{15}\right) = P(Z \ge 2) = 0.0228$$

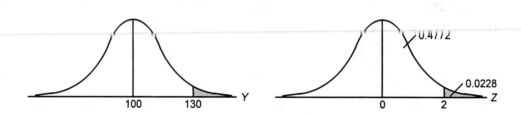

$$P(70 < Y < 130) = P(-2 < Z < 2) = 0.9544$$

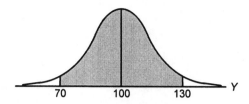 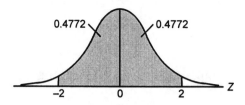

5.58 Let Y represent the hourly wage rate earned by a worker in a clothing factory. We are assuming that the random variable Y is normally distributed with $\mu = 5.10$ and $\sigma = 0.40$.

a. $P(Y > 5.40) = P\left(\dfrac{Y - \mu}{\sigma} > \dfrac{5.40 - 5.10}{0.40}\right) = P(Z > 0.75) = 0.2266$

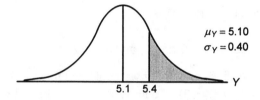 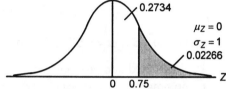

b. $P(4.70 < Y < 5.50) = P\left(\dfrac{4.70 - 5.10}{0.40} < \dfrac{Y - \mu}{\sigma} < \dfrac{5.50 - 5.10}{0.40}\right) = P(-1 < Z < 1) = 0.6826$

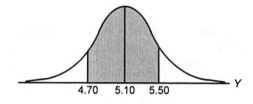 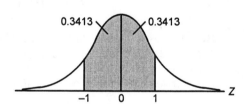

c. $P(Y > 3.90) = P\left(\dfrac{Y - \mu}{\sigma} > \dfrac{3.90 - 5.10}{0.40}\right) = P(Z > -3) = 0.9987$

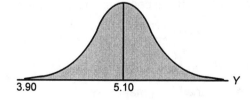 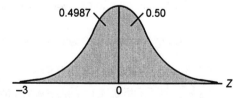

Supplementary Exercises

5.61 Y is the number of Medicaid claims in a particular city that are fraudulent. Each claim may be regarded as a trial, with success defined as a fraudulent claim. (Here's a case where "success" is an arbitrary choice!) The probability of success should be 0.05 on each trial, and there's no reason to suppose that the trials are dependent. Therefore, Y has a binomial probability distribution with $n = 50$ and $\pi = 0.05$.

a. To find $P(Y \le 1)$, use Appendix Table 1. Note, that

$$P(Y \le 1) = P(Y = 0) + P(Y = 1)$$

Identify $n = 50$, $\pi = 0.05$; add the entries for $y = 0$ and 1, and get 0.2794.

b. We could add probabilities for $y = 4$, 5, 6, and up; or we could find the complementary probability. $P(Y \ge 4) = 1 - P(Y \le 3)$. To find $P(Y \le 3)$, as in part **a.**, identify $n = 50$, $\pi = 0.05$ but now add the entries for $y = 0$, 1, 2, 3, and get 0.7604
Therefore,

$$P(Y \ge 4) = 1 - 0.7604 = 0.2396$$

5.62   **a.** This problem only requires counting, so the question is whether to consider sequences (permutations) or subsets (combinations). There is no need to consider the order in which the 14 fringe managers are selected. Therefore, the combination rule applies with $r = 30$ people, of whom $k = 14$ are to be chosen.

$$\binom{r}{k} = \binom{30}{14} = \frac{30!}{14!16!} = 145,422,675$$

b. To obtain all choices with 5 women and 9 men, we can combine any of the $\binom{6}{5}$ choices of 5 from the 6 women with any of the $\binom{24}{9}$ choices of 9 from the 24 men. Since any choice of women can be matched with any choice of men, there are $\binom{6}{5}\binom{24}{9} = 7,845,024$ possible ways to choose the fringe managers.

c. Use the classical interpretation of probability, because all possible choices are assumed to be equally likely. Divide the number of choices satisfying the definition by the total number of choices. It may be easier to see the logic if we add the probabilities for exactly 5 and exactly 6. Note that 6 is the maximum possible, because there are only 6 women.

$$P(Y \geq 5) = P(Y = 5) + P(Y = 6) = \frac{\binom{6}{5}\binom{24}{6}}{\binom{30}{14}} - \frac{\binom{6}{6}\binom{24}{8}}{\binom{30}{14}} = \frac{8{,}580{,}495}{145{,}422{,}675} = 0.059$$

5.63 Y is the number of lost time industrial accidents in a 10-day period. Y has a Poisson distribution with $\mu = $ (rate per day)(number of days) $= 0.12(10) = 1.2$.

a. Appendix Table 2 is used to find the following probabilities. Enter the column where $\mu = 12$.

$$P(Y = 1) = 0.3614$$
$$P(Y \leq 1) = P(Y = 0) + P(Y = 1) = 0.6626$$

b. For the Poisson distribution, the standard deviation is the square root of the expected value.

$$E(Y) = \mu = 12$$
$$\sigma_Y = \sqrt{\mu} = \sqrt{12} = 1095$$

5.64   By assumption, Y, the demand for 5 pound sacks of flour in a particular week, is approximately normally distributed with $\mu = 72$ and $\sigma = 1.6$.

a. To calculate normal probabilities, convert to Z and use Appendix Table 3, drawing the appropriate sketch of a normal curve.

$$P(Y \leq 72.8) = P\left(\frac{Y - \mu}{\sigma} \leq \frac{72.8 - 72}{1.6}\right) = P(Z \leq 0.5) = 0.6915$$

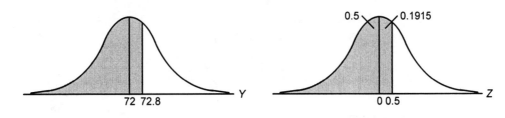

$$P(71.2 \leq Z \leq 72.8) = P\left(\frac{71.2-72}{1.6} \leq \frac{Y-\mu}{\sigma} \leq \frac{72.8-72}{1.6}\right) = P(-0.5 \leq Z \leq 0.5)$$
$$= 0.1915 + 0.1915 = 0.3830$$

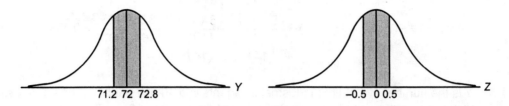

b. Again, convert to Z and draw a sketch.

$$P(Y \geq 74) = P\left(\frac{Y-\mu}{\sigma} \geq \frac{74-72}{1.6}\right) = P(Z \geq 1.25) = 0.1056$$

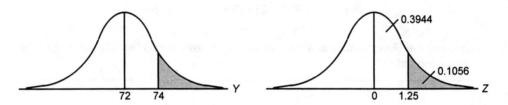

c. $P(Y > k) = P\left(\frac{Y-\mu}{\sigma} > \frac{k-72}{1.6}\right) = P(Z > z) = 0.01$

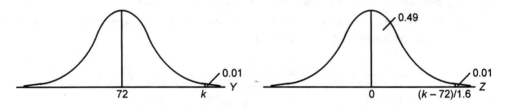

From Table 3, $P(Z > 2.33) = 0.01$
Equating equivalent quantities, we have that

$$z = \frac{k-72}{1.6} = 2.33$$

Solving for k, we have that

$$k = 72 + 2.33(1.6) = 75.7$$

60

5.65　　**a.**　This is the now-familiar normal probability calculation.

$$P(Y > 73.0) = P\left(Z > \frac{73.0 - 72.0}{1.6}\right) = P(Z > 0.625)$$

The table entry for 0.62 is 0.2324; for 0.63 it is 0.2357. If we take a value halfway between these two, we get a rough table value for 0.625 of 0.2341. Thus, approximately,

$$P(Y > 73.0) = 0.5 - 0.2341 = 0.2659$$

b.　Each week can be thought of as a trial. Apply the binomial probability distribution to find the probability that the demand exceeds 73 cases in exactly 3 of 4 consecutive weeks with parameters,

$$n = 4 \qquad \pi = 0.2659 \text{ (from part a.)} \qquad y = 3$$

Therefore,

$$P(Y = 3) = \binom{4}{3}(0.2659)^3(0.7341)^1 = 0.055$$

5.68　　Let Y be the number of errors in transmission. Y has a Poisson distribution with

$$\mu = \left(\frac{1}{5,000}\right)(25,000) = 5$$

and

$$\sigma_Y = \sqrt{\sigma} = \sqrt{5} = 2.236$$

a.　The probability that the device will be accepted is the same as the probability that the number of errors in transmission is less than 8 (not including 8). Therefore, using Appendix Table 2, with $\mu = 5.0$,

$$P(Y < 8) = P(Y = 7) + P(Y = 6) + \cdots + P(Y = 0) = 0.8666$$

b.　We might expect the independence assumption not to hold. An error in transmission may indicate noisy lines or other problems.

5.69　　Now we assume that the modem has a mean error rate of 1 per 2,500 words so that

$$\mu = \left(\frac{1}{2,500}\right)(25,000) = 10$$

Then using Appendix Table 2 with $\mu = 10.0$,

$$P(Y < 8) = P(Y = 7) + \cdots + P(Y = 0) = 0.2203$$

5.75 The correct probability distribution should be binomial. All the assumptions seem reasonable, unless one worries about improbabilities such as identical twins both having the defect. Here $n = 5{,}000$ and $\pi = 0.0001$. Minitab computed the probability of 0 babies with the defect.

```
MTB > PDF 0;
SUBC>    Binomial 5000 .0001.

Binomial with n = 5000 and p = 0.000100000

        x          P( X = x)
      0.00           0.6065
```

Therefore the probability of at least one baby with the defect is $1 - 0.6065 = 0.3935$.

5.81 **a.** The expected value is the rate per day times the number of days.

$$\mu_Y = (0.5 \text{ per day})(7 \text{ days}) = 3.5$$

b. We assume that Poisson probabilities apply with $\mu_Y = 3.5$. From Appendix Table 2

$$P(Y \geq 4) = P_Y(4) + P_Y(5) + \cdots = 0.1888 + 0.1322 + \cdots + 0.0001 = 0.4634$$

c. The Poisson assumptions are nonclumping (we don't have several homes involved in one fire) and independence (the occurrence of one fire doesn't change the probability of having another fire at a later time). One could image a mad arsonist as a violation of the assumptions! More plausibly, a fire might cause others to be more careful for a while, reducing the probability of another fire; this would violate independence.

Chapter 6

Random Sampling and Sampling Distributions

6.1 Random Sampling

6.2 There is a selection bias. Those who are at home between 3:00 and 6:00 are presumably more likely to have school-age or younger children. Therefore the process is not a good approximation to random sampling.

6.5 There is a size bias. A purchaser who holds tickets to many seats is more likely to be sampled than a purchaser who holds tickets to only one or two seats.

6.2 Sample Statistics and Sampling Distributions

6.7 To find the population mean, we use the definition

$$\mu = E(Y) = \sum_{\text{all } y} y P_Y(y) = 4(0.50) + 8(0.30) + 12(0.15) + 16(0.05)$$
$$= 2.0 + 2.4 + 1.8 + 0.8 = 7.0$$

Use the definition (or the shortcut method, if preferred) to find the variance and standard deviation.

$$\text{Var}(Y) = (4-7)^2(0.50) + (8-7)^2(0.30) + (12-7)^2(0.15) + (16-7)^2(0.05) = 12.60$$

so

$$\sigma_Y = \sqrt{12.60} = 3.5496$$

To find the expected value of $\overline{Y}$, we use the definition

$$E(\overline{Y}) = \sum_{\text{all } \overline{y}} \overline{y} P_{\overline{Y}}(\overline{y}) = 4(0.0039) + 4.5(0.0188) + \cdots + 16(0)$$
$$= 0.0156 + 0.0846 + \cdots + 0 \approx 7$$

To find the variance of $\overline{Y}$, we use the definition

$$\sigma_{\overline{Y}}^2 = \sum_{\text{all } \overline{y}} (\overline{y} - \mu_{\overline{Y}})^2 P(\overline{y}) \qquad \text{where } \mu_{\overline{Y}} = E(\overline{Y})$$
$$= (4-7)^2(0.0039) + (4.5-7)^2(0.0188) + \cdots + (16-7)^2(0)$$
$$= 1.575$$

so

$$\sigma_{\overline{Y}} = \sqrt{1.575} = 1.2550$$

The expected value of the sample mean is 7, equal to the population mean. The standard deviation of the sample mean (standard error of the mean) is

$$\sigma_{\overline{Y}} = \frac{\sigma}{\sqrt{n}} = \frac{3.5496}{\sqrt{8}} = 1.2550$$

6.3   Sampling Distributions for Means and Sums

6.10   **a.** Let $T = \text{total}$, so $T = n\overline{Y}$. $\mu_T = n\mu = 10(327) = 3270$. $\sigma_T^2 = n^2\sigma_{\overline{Y}}^2 = n\sigma^2$, and
$\sigma_T = \sqrt{n}\sigma = \sqrt{10}(34) = 107.52$

b. $P(3150 < T < 3390) = P\left(\dfrac{3150 - 3270}{107.52} < Z < \dfrac{3390 - 3270}{107.52}\right) = P(-1.12 < Z < 1.12)$
$= 2(0.3686) = 0.7372$

c. $\mu_{\overline{Y}} = \mu = 327$. $\sigma_{\overline{Y}} = \dfrac{\sigma}{\sqrt{n}} = \dfrac{34}{\sqrt{10}} = 10.75$.

$$P\left(314 < \overline{Y} < 339\right) = P\left(\dfrac{314 - 327}{10.75} < Z < \dfrac{339 - 327}{10.75}\right) = P(-1.21 < Z < 1.12)$$
$$= 0.3686 + 0.3869 = 0.7555$$

6.11 $P\left(327 - \kappa \le \overline{Y} \le 327 + \kappa\right) \approx 0.95 \Rightarrow P\left(\dfrac{327 - \kappa - 327}{10.75} < Z < \dfrac{327 + \kappa - 327}{10.75}\right)$

$$\approx 0.95 \Rightarrow P\left(\dfrac{-\kappa}{10.75} < Z < \dfrac{\kappa}{10.75}\right) \approx 0.95$$

The point on the Z curve with 0.95 between is -1.96 and 1.96. So $\kappa/10.75 = 1.96$, or $\kappa = 21.07$. Therefore, the range is $327 \pm 21.07 = (305.93,\ 348.07)$.

6.16 **a.** This question deals with an individual observation, as opposed to a sample mean. In effect, the sample size is 1. We are assuming that normal probabilities apply, so we convert to z scores and use Appendix Table 3.

$$P(\text{Demand} > 170) = P\left(Z > \dfrac{170 - 148}{21}\right) = P(Z > 1.05)$$

The table entry (area under the curve) for $z = 1.05$ is 0.3531. Because we want the area beyond $z = 1.05$, the desired probability is $0.5 - 0.3531$, or 0.1469.

b. Now we are dealing with a sample mean, based on $n = 12$ Saturdays. Again, normal probabilities apply, but we must use the standard error of the sample mean, namely $\sigma/\sqrt{n} = 21/\sqrt{12} = 6.06$. Therefore,

$$P(\text{sample mean} < 135 \text{ or sample mean} > 155) = 1 - P\left(135 \le \overline{Y} \le 155\right)$$
$$= P(-2.14 \le Z \le 1.15)$$

The areas in Appendix Table 3 for $z = 2.14$ and 1.15 are 0.4838 and 0.3749, respectively, so $P(-2.14 \le Z \le 1.15) = 0.4838 + 0.3749 = 0.8587$, and the desired probability is $1 - 0.8587 = 0.1413$.

6.17 **a.** According to the argument, most of the demand values will be near the mean, with a few much smaller values. This is the pattern of left-skewed data.

b. The Central Limit Theorem effect does some good, even for a sample mean of as few as 12 observations. But it does no good at all for a sample mean of 1 observation, that is, for an individual observation. Because the answer in part **a.** is based on an individual observation, and the answer in **b.** is based on a sample mean of several observations, the **a.** answer will be less accurate if the population isn't normal.

6.20   **a.** Again, we are dealing with a single observation ($n = 1$, if you wish), not a sample mean.

$$P(Y > 80) = P\left(Z > \frac{80 - 71}{15}\right) = P(Z > 0.60)$$

We seek the area beyond $z = 0.60$. The tabled area between 0 and 0.60 is 0.2257, so the desired probability is $0.5 - 0.2257 = 0.2743$.

b. Now we're concerned about a sample mean from a sample of $n = 10$ observations. The required standard error is $15/\sqrt{10} = 4.74$.

$$P(\overline{Y} > 80) = P\left(Z > \frac{80 - 71}{4.74}\right) = P(Z > 1.90) = 0.5 - 0.4713 = 0.0287$$

6.21   In this exercise, we're assuming a severely skewed distribution of individual values, and a sample size of only 10. Standard guidelines for the applicability of the Central Limit Theorem indicate that a sample size of 10 isn't adequate to assume normal probabilities, given a skewed population. Note that the probability is a one-tailed probability, so that there is no compensating underestimate of one tail, overestimate of the other. We would not think that the approximation was very accurate.

6.4   Checking Normality

6.22   **a.** The histogram has a peak at about 795. There seems to be a somewhat longer tail on the left (low) side of the data. We'd say there was a slight left skewness in the histogram.

b. The actual values are shown along the horizontal axis, and marked "Cleared." On the vertical axis, the labels show "Probability," and are presumably the theoretical values. We think we see a slight upward curve as we go to the right in the plot. Notice that the points at both ends of the scale are above the reference line. When the data are along the horizontal axis of a normal plot, an upward-curving plot indicates left skewness. The pattern isn't hugely obvious, certainly, but we think it shows some degree of left skew.

6.23   As with any normal plot, the first thing to check is which axis is which. In this plot, the data are along the *vertical* axis, and the theoretical, normal scores are along the horizontal axis. In the plot, there is a very clear curve, getting steeper as we go to the right. According to the text, when the data are on the vertical axis, such a curve indicates right skewness. That makes sense in context. Most automobile collision claims will be modest, but a few will be very large; we should see right skewness in such data.

6.24   The data are shown along the horizontal axis. Most of the points fall along the line quite nicely. There is perhaps a slight curve down at the high end of the data. We would call these scores nearly normally distributed, but wouldn't argue too much if you called it slightly right skewed.

6.27   There is a definite curve in the plot, with the slope getting flatter as we move to the right. The data are along the horizontal axis, labelled as "Maxof10." Such a plot indicates clear right skewness. The distribution is *not* close to normal.

6.28   **a.** This plot looks almost the same as the plot in the previous exercise. If anything, we think that there's even more of a curve. The distribution is not close to normal, but definitely right skewed.

b. As we noted, the plot seemed to get a bit *more* skewed as the sample size increased. If the Central Limit Theorem worked for sample maxima, the plot should becom *less* skewed as n increases. Therefore, it appears that the Central Limit Theorem does not work for this statistic.

Supplementary Exercises

6.29   **a.** We are sampling from a population having 20% red beads, and the sample size is 50. (It doesn't matter whether we regard the sample as being taken with or without replacement.) The expected number of red beads in the sample should be 20% of 50, or 10.

b. This is a somewhat subtle question. We are implicitly assuming that the probability of a red bead being sampled is the same as its proportion in the sample, namely 0.20. This is only true if the paddle sample is completely random, so that the probability that any particular bead is chosen is the same, regardless of its color. In other words, we are assuming that the paddle mechanism is unbiased in sampling. The assumption might be wrong. For instance, the red paint on those beads might make them larger, less likely to fit in the holes, and less likely to be sampled. Or the paint might make them stickier, and more likely to be sampled.

6.30   We assume that the deviation from the assumed 10 is more than random, given that the result is based on thousands of samples. If so, the expected value is not 10, but (close to) 9.4, so that the sampling is mildly biased against drawing red beads.

6.35   **a.** The expected value of the sample mean is the population expected value for each item. In this case, the expected value per disk is 2.13 K. The expected value of the mean bad-sector volume in a pack is also 2.13.

b. The standard deviation of the mean, assuming random sampling and independence, is the standard deviation per item divided by the square root of the number of items. Therefore

$$\sigma_{\overline{Y}} = \frac{\sigma}{\sqrt{100}} = \frac{0.83}{10} = 0.083$$

6.36   The most critical assumption is that the assembly of disks is random, so far as bad sectors are concerned. In particular, we assumed that the size of bad sectors was independent from one disk in the pack to another. This assumption is important only for the standard deviation calculation. For example, if the retailer compensated for an exceptionally bad disk by inserting an exceptionally good disk into the same pack, the variability of the total would be reduced. Alternatively, if the disks were packed in the order they were produced, and if bad sectors ran in streaks, the variability of the total bad-sector volume would be increased over the standard deviation found in the previous exercise.

6.37   We would expect this histogram to have a normal distribution shape. The Central Limit Theorem applies to means. Given that the mean bad-sector volume is the average of 100 independent (more or less) terms, the sample size is more than adequate to assume that the Central Limit Theorem effect will occur.

6.45   **a.** The given information states probabilities for sample means outside various values. To find the exact probability that $\overline{Y}$ will be within 1 standard error of μ, we can use the complements principle. The probability that the sample mean will be *within* one standard deviation of the mean is 1 minus the probability that it is outside that range.

$$P\left(\mu - \sigma_{\overline{Y}} < \overline{Y} < \mu + \sigma_{\overline{Y}}\right) = 1 - \left[P\left(\overline{Y} < \mu - \sigma_{\overline{Y}}\right) + P\left(\overline{Y} > \mu + \sigma_{\overline{Y}}\right)\right]$$

Therefore, for

$$n = 2$$

$$P\left(\mu - \sigma_{\overline{Y}} < \overline{Y} < \mu + \sigma_{\overline{Y}}\right) = 1 - [0.0888 + 0.0888] = 0.8224$$

$n = 4$

$$P\left(\mu - \sigma_{\overline{Y}} < \overline{Y} < \mu + \sigma_{\overline{Y}}\right) = 1 - [0.0899 + 0.0899] = 0.8202$$

$n = 8$

$$P\left(\mu - \sigma_{\overline{Y}} < \overline{Y} < \mu + \sigma_{\overline{Y}}\right) = 1 - [0.1456 + 0.1456] = 0.7088$$

b. If the normal approximation is used to find the probability, we would have

$$P\left(\mu - \sigma_{\overline{Y}} < \overline{Y} < \mu + \sigma_{\overline{Y}}\right) \approx 0.6826$$

The normal approximation is not good for $n = 2$ and $n = 4$ but is fairly good for $n = 8$. The normal approximation gets closer to the actual probabilities as the sample size increases.

c. Again, we can use the complements idea to find the probability that the sample mean will be within two standard errors of the population mean.

$$P\left(\mu - 2\sigma_{\overline{Y}} < \overline{Y} < \mu + 2\sigma_{\overline{Y}}\right) = 1 - \left[P\left(\overline{Y} < \mu - 2\sigma_{\overline{Y}}\right) + P\left(\overline{Y} > \mu + 2\sigma_{\overline{Y}}\right)\right]$$

Therefore, for

$n = 2$

$$P\left(\mu - 2\sigma_{\overline{Y}} < \overline{Y} < \mu + 2\sigma_{\overline{Y}}\right) = 1 - [0.0388 + 0.0388] = 0.9224$$

$n = 4$

$$P\left(\mu - 2\sigma_{\overline{Y}} < \overline{Y} < \mu + 2\sigma_{\overline{Y}}\right) = 1 - [0.0488 + 0.0488] = 0.9024$$

$n = 8$

$$P\left(\mu - 2\sigma_{\overline{Y}} < \overline{Y} < \mu + 2\sigma_{\overline{Y}}\right) = 1 - [0.0333 + 0.0333] = 0.9334$$

If the normal approximation is used to find the probability, we would have

$$P\left(\mu - 2\sigma_{\overline{Y}} < \overline{Y} < \mu + 2\sigma_{\overline{Y}}\right) = 0.95$$

The normal approximation is fairly close even for the $n = 2$ sample, but again best for $n = 8$.

6.46   **a.** A histogram or stem-and-leaf display can be obtained from almost any package. See the documentation or help file of the program you're using to find out how. Different programs will use slightly different methods for presenting a histogram. Here is a Minitab version.

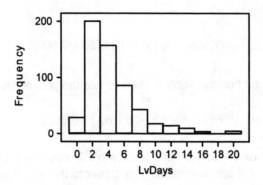

The data are right-skewed, not normally distributed.

b. Here is what Minitab yielded.

```
MTB > describe 'LvDays'

Variable        N       Mean    Median    TrMean    StDev    SEMean
LvDays          533     3.623   3.000     3.334     2.904    0.126

Variable        Min     Max       Q1        Q3
LvDays          0.000   19.000    2.000     5.000
```

The mean is 3.623; your program may give a different number of decimal places. The standard deviation is 2.904. If the program regards the data as a sample, it will divide by $n-1 = 532$, not by $N = 533$. The numerical effect of this distinction is tiny, and not worth worrying about.

c. Because the samples are taken at random, the answer you get will differ from ours.

d. A stem-and-leaf display of the means we got using Minitab is shown here.

```
MTB > Stem and Leaf of 'Means'

Character Stem-and-Leaf Display

Stem-and-leaf of Means      N  = 25
Leaf Unit = 0.10


     2      2 77
     2      2
     3      3 1
     7      3 2233
    10      3 445
    (5)     3 66677
    10      3 8
     9      4 0011
     5      4 3
     4      4 44
     2      4 7
     1      4 8
```

Again, your answer will differ slightly from ours because the samples are taken randomly. The shape will not be the same as the population shape. Because of the Central Limit Theorem, the distribution of sample means will be much more nearly normal than the population distribution. Because n is only 20, you may still see some slight right skewness.

6.47   Again, your answer will differ slightly from ours due to random sampling. We obtained a standard deviation of 0.565. The standard deviation of the means is an approximation to the standard error of the sample mean. Theoretically, the value should be 2.904 divided by the square root of 20, or 0.65. We happened to come out a bit too low.

Review Exercises—Chapters 4–6

R.26   **a.** To find the probabilities, we must use appropriate x values in the defining formula for $P_x(x)$. For example, for $x = 0$, $P_x(0) = (0 + 1)/66 = 1/66$. For $x = 1$, $P_x(1) = (1 + 1)/66 = 2/66$. Continue in this fashion up to $P_x(10) = (10 + 1)/66 = 11/66$.

x	0	1	2	3	4	5	6	7	8	9	10
$P_x(x)$	1/66	2/66	3/66	4/66	5/66	6/66	7/66	8/66	9/66	10/66	11/66

b. We must use the definition of expected value, because this is not one of the special probability distributions with a shortcut calculation for the expected value. To find the mean (expected value) of a random variable X, weight each value by its probability and sum.

$$\mu_x = \sum x P_x(x) = 0\left(\frac{0+1}{66}\right)+\cdots+10\left(\frac{10+1}{66}\right) = 6.66667$$

The easiest way to find the standard deviation is to use the shortcut method for the variance

$$\sigma_x^2 = \sum x^2 f_x(x) - \mu_x^2 = 0^2\left(\frac{0+1}{66}\right)+\cdots+10^2\left(\frac{10+1}{66}\right) - (6.66667)^2$$

$$= 51.66667 - 44.44444 = 7.22223$$

$$\sigma_x = \sqrt{7.22223} = 2.687$$

c. To have enough parts in stock is to have $X \geq 3$. We can either add the probabilities for $x = 3$, 4, ..., 10, or use the complements idea.

$$P(X \geq 3) = \sum_{i=3}^{10} P_x(x) = \sum_{i=3}^{10} \frac{(x+1)}{66} = \frac{60}{66} = 0.909$$

R.27   **a.** In effect, we assume that the four parts may be regarded as a random sample of size $n = 4$. Then we can use the basic properties of sampling distributions. The exepcted value of the sample mean is the same as the population mean, in this case 6.66667. The standard error of the sample mean is the population standard deviation, here $\sqrt{7.22223}$ divided by the square root of $n = 4$. The variance is the square of the standard error, and therefore is the population variance divided by the sample size.

$$\text{Var}(\overline{Y}) = \frac{\text{Var}(Y)}{n} = \frac{7.22223}{4} = 1.8056$$

b. We assumed independence in calculating the variance. This is a basic assumption of random sampling. The assumption says that there is no carryover effect from one observation to the next. The assumption should be valid if the data are truly based on a random sample. This assumption is important in finding the variance, but irrelevant in finding the mean.

R.28   We are dealing with the probability (sampling) distribution of $\overline{Y}$, the mean of 200 random variables. We know that

$$\mu_{\overline{Y}} = \mu = 6.66667$$

and

$$\sigma_{\overline{Y}} = \frac{\sigma}{\sqrt{n}} = \frac{\sqrt{7.22223}}{\sqrt{200}} = 0.1900$$

The Central Limit Theorem indicates that normal probabilities apply. Therefore

$$P(\overline{Y} > 7) = P\left(Z > \frac{7 - 6.66667}{0.1900}\right) = P(Z > 1.75) = 0.5000 - 0.4599 = 0.0401$$

The normal approximation is based on a sample size of 200. The population distribution given in Exercise **R.26** is skewed, but the sample size of 200 is large enough that the Central Limit approximation should be highly accurate.

R.37   Each customer should be thought of as one trial. Define a success to be a customer who eventually purchases a home through the realtor. We assume that binomial probabilities apply, with $n = 16$ and $\pi = 0.45$. To find $P(Y \leq 3)$ we must add the probabilities of $y = 0$, 1, 2, and 3 found in Appendix Table 1.

$$P(Y \leq 3) = P_Y(0) + P_Y(1) + P_Y(2) + P_Y(3) = 0.0001 + 0.0009 + 0.0056 + 0.0215 = 0.0281$$

R.38   We assumed that each customer either does or does not purchase a home, that the purchase probability is the same for all customers, that whether or not one customer purchases a home has no effect on the probability of other customers purchasing, and that the sample size of 16 was fixed with the order of purchases irrelevant. All these assumptions seem very plausible.
        One might argue that the probability of purchase varies because customers differ in income, taste, and other factors. But these characteristics are themselves random; a different sample of 16 customers would differ on these factors. There is no particular reason why customer number 2 (whoever that might be) is more or less likely to purchase than customer number 1.

R.43   There are no specific trials here, but rather a continuing stream of special handling orders coming in over time. We assume that Poisson probabilities apply. The expected number $\mu_Y = (1.6/\text{day})(5 \text{ days}) = 8.0$. We must add probabilities for $y = 10$, 11, ... in the 8.0 column of the Poisson table, Appendix Table 2

$$P(Y \geq 10) = 0.0993 + 0.0722 + 0.0481 + 0.0296 + 0.0169 + 0.0090$$
$$+0.0045 + 0.0021 + 0.0009 + 0.0004 + 0.0002 + 0.0001$$
$$= 0.2833$$

R.44   The Poisson assumptions are that the wholesaler doesn't receive two or more special-handling orders simultaneously (nonclumping) and that what happens in one time period doesn't affect the probabilities for another time period (independence). There might be clumps of special-handling orders at the release of a new product. Apart from that, barring some farfetched scenarios such as several customers banding together to send in special-handling orders, the assumptions seem sensible.

Chapter 7

Estimation

7.1 Point Estimators

7.3 **a.** All three estimators—the mean, the median, and the 20% trimmed mean—appear unbiased. The averages are all very close to 100. This result should hold, because the simulation used a symmetric (in fact normal) population.

b. The mean has the smallest variance, and therefore will have the smallest standard error. The sample mean appears most efficient. This result follows *given* a normal population. It may not hold for other populations.

7.4 **a.** The mean is 236.4 and the median is 234.5.

b. The 20% trimmed mean is 233.875.

7.5 **a.**

Col. Inch Stem and Leaf

Stem	Leaf	Count
36	9	1
34		
32		
30		
28	0	1
26	2 3 1	3
24	0 1 3 5 9 1 4 7	8
22	1 3 5 8 8 0 4 5 7	9
20	4 0 6 8	4
18	5 3 9	3
16	1	1

Multiply Stem.Leaf by 10

b. The outlier suggests that the median is probably more efficient than the mean in this situation.

7.2　Interval Estimation of a Mean, Known Standard Deviation

7.9　To calculate a confidence interval, we need the sample mean, sample size, and an assumed standard deviation for the data. The data from the random sample of $n = 20$ VP's yielded $\bar{y} = 23.985$ and we assume that $\sigma = 4$.

a.　The general 95% confidence interval for μ is

$$\bar{y} - z_{\alpha/2}\frac{\sigma}{\sqrt{n}} \leq \mu \leq \bar{y} + z_{\alpha/2}\frac{\sigma}{\sqrt{n}}$$

In this case, $\alpha = 0.05$; the required table value is the familiar 1.96. $z_{\alpha/2} = z_{0.025} = 1.96$. Therefore,

$$23.985 - \frac{1.96(4)}{\sqrt{20}} \leq \mu \leq 23.985 + \frac{1.96(4)}{\sqrt{20}}$$
$$23.985 - 1.7531 \leq \mu \leq 23.985 + 1.7531$$
$$22.2319 \leq \mu \leq 25.7381$$

rounding:

$$22.23 \leq \mu \leq 25.74 \text{ with 95% confidence}$$

b.　The other information we need for a 99% confidence interval for μ is the appropriate z table value. For 99% confidence, $\alpha = 0.01$; this probability is divided, 0.005 in each tail of the normal curve. We need an area of 0.495 between 0 and the required z value. The closest value in the table is $z = 2.58$. The confidence interval in general is

$$\bar{y} - z_{\alpha/2}\frac{\sigma}{\sqrt{n}} \leq \mu \leq \bar{y} + z_{\alpha/2}\frac{\sigma}{\sqrt{n}}$$

In this case, $z_{\alpha/2} = z_{0.005} = 2.58$. Therefore

$$23.985 - \frac{2.58(4)}{\sqrt{20}} \le \mu \le 23.985 + \frac{2.58(4)}{\sqrt{20}}$$

$$21.68 \le \mu \le 26.29$$

with 99% confidence

7.10 We would expect that 95% of all confidence intervals calculated in this manner would include the parameter μ. "In this manner" means that if we sampled again and again (20 persons each time) and calculated a sample mean and a 95% C.I. for the true mean each time, we would expect that 95% of them include the population mean, μ.

7.11 Any data plot (histogram, stem-and-leaf, box plot, or normal plot) indicates that the data are nearly normal. Therefore, the population should be nearly normal, so the distribution of the sample mean will also be very nearly normal.

7.13 The data from the sample of 36 items yielded a sample average "shrinkage" of 5.8% ($\overline{y} = 5.8$) with a standard deviation of 4.2% ($s = 0.4$). We must assume (probably wrongly) that the population standard deviation exactly equals the sample standard deviation.

An approximate 95% confidence interval for μ is

$$\overline{y} - z_{\alpha/2} \frac{s}{\sqrt{n}} \le \mu \le \overline{y} + z_{\alpha/2} \frac{s}{\sqrt{n}}$$

where

$$\alpha = 0.05; \quad z_{\alpha/2} = z_{0.025} = 1.96$$

Therefore,

$$5.8 - \frac{1.96(4.2)}{\sqrt{36}} \le \mu \le 5.8 + \frac{1.96(4.2)}{\sqrt{36}}$$

$$5.8 - 1.372 \le \mu \le 5.8 + 1.372$$

$$4.4280 \le \mu \le 7.172$$

Rounding:

$$4.43\% \le \mu \le 7.17\%$$

with 95% confidence

7.14 At best, maybe. The distribution of Y may not be normal since shrinkage will not take on negative values; the standard deviation is so large that the Empirical Rule must fail. However the CLT tells us that for n large and Y not severely skewed, the distribution of $\overline{Y}$ will be approximately normal. It isn't clear that $n = 36$ is enough to "deskew" the sampling distribution. We would think that the approximation is mediocre.

7.3   Confidence Intervals for a Proportion

7.20   The sample proportion of individuals who were aware of a certain product is $\hat{\pi} = 84/125 = 0.672$. A 90% confidence interval for π is

$$\hat{\pi} - z_{\alpha/2}\sqrt{\frac{\hat{\pi}(1-\hat{\pi})}{n}} \le \pi \le \hat{\pi} + z_{\alpha/2}\sqrt{\frac{\hat{\pi}(1-\hat{\pi})}{n}}$$

where in this case

$$\alpha = 0.10; \quad z_{\alpha/2} = z_{0.05} = 1.645$$

Therefore,

$$0.672 - 1.645\sqrt{\frac{0.672(0.328)}{125}} \le \pi \le 0.672 + 1.645\sqrt{\frac{0.672(0.328)}{125}}$$
$$0.672 - 0.0691 \le \pi \le 0.672 + 0.0691$$
$$0.6029 \le \pi \le 0.7411$$

Rounding

$$0.60 \le \pi \le 0.74$$

with 90% confidence

7.21   Yes. The sample size (125) is large enough and $n\hat{\pi}$ and $n(1-\hat{\pi})$ are both greater than 5 $(n\hat{\pi} = 125(0.672) = 125(0.572) = 84,\ n(1-\hat{\pi}) = 41)$ so that the normal approximation to the binomial is accurate.

7.4　How Large a Sample Is Needed?

7.24　Desired: 95% confidence interval with width of 0.3σ so that

$$2E = 0.3\sigma \quad E = 0.15\sigma \quad \alpha = 0.05 \quad z_{\alpha/2} = 1.96$$

Therefore,

$$n = \frac{(1.96)^2 \sigma^2}{(0.15\sigma)^2} = \frac{1.96^2}{0.15^2} \approx 171$$

Desired: 95% confidence interval with width of 0.4σ so that

$$2E = 0.4\sigma \quad E = 0.2\sigma$$

Therefore,

$$n = \frac{(1.96)^2 \sigma^2}{(0.2\sigma)^2} = \frac{1.96^2}{0.2^2} \approx 96$$

7.28　**a.** To find a 95% confidence interval with a width of 0.02, we need the allowable error E, the table value corresponding to 95% confidence, and a value for π.

$$2E = 0.02 \quad E = 0.01 \quad \alpha = 0.05 \quad z_{\alpha/2} = 1.96$$

Therefore, using the conservative estimate, $\hat{\pi} = 0.5$,

$$n = \frac{(1.96)^2 (0.5)(0.5)}{(0.01)^2} \approx 9,604$$

b. If we assume that the proportion of imperfect boxes is $\hat{\pi} = 0.005$, the required sample size would be

$$n = \frac{(1.96)^2 (0.005)(0.995)}{(0.01)^2} \approx 191$$

However, the more crucial value for π is the value closest to the conservative 0.50; in this case, that is $\pi = 0.08$. If it were assumed that the proportion of imperfect boxes is $\hat{\pi} = 0.08$, the required sample size would be

$$n = \frac{(1.96)^2(0.08)(0.92)}{(0.01)^2} \approx 2{,}827$$

The sample size need only be 2,827 (as compared to 9,604 in the previous part) for a width no larger than 0.02 given the assumption that the proportion of imperfect boxes is somewhere in the interval $0.005 \leq \pi \leq 0.08$.

7.5   The *t* Distribution

7.31   **a.** Simulated and theoretical relative frequencies are tabulated below:

Range	Actual Frequencies	Relative Frequencies	Theoretical Relative Frequencies (df = 3)
$t < -2.353$	44	0.0400	$P(t < -2.353) = 0.05$
$-2.353 < t < -1.638$	59	0.0536	$P(-2.353 < t < -1.638) = 0.05$
$-1.638 < t < 1.638$	896	0.8145	$P(-1.638 < t < 1.638) = 0.80$
$1.638 < t < 2.353$	47	0.0427	$P(1.638 < t < 2.353) = 0.05$
$2.353 < t$	54	0.0491	$P(2.353 < t) = 0.05$
	1,100	1	1

b. There is no evidence of a systematic departure from the theoretical frequencies (the actual relative frequencies are all reasonably close to their respective theoretical frequencies).

7.6   Confidence Intervals for the *t* Distribution

7.32   **a.** The 99% confidence interval is shown under the headings LO 99% CI and UP 99% CI. It is

$$39.841 \leq \mu \leq 73.898$$

b. The boxplot has a very short lower tail and short lower half of the box. The upper box and upper tail are much longer. This pattern indicates right skewness of the data. In addition, there are two outliers on the high end. Because the sample data shape tends to reflect the population data shape, we tend to believe that the population is right skewed.

c. The interval is extremely wide. It ranges from about 40 days to about 74. The width is a full month, plus. That is not a very accurate estimate of the average age.

7.33   We want a 90% confidence interval with a width of 6 days. Start by assuming that n will be large enough to ignore the difference between *t* and *z* tables. We need an allowable ± *E*, a table value, and a standard deviation to do the calculations.

$$\alpha = 0.1 \quad z_{\alpha/2} = z_{0.05} = 1.645$$

$$2E = 6 \quad E = 3 \quad s = 28.97$$

Therefore,

$$n = \frac{\left(z_{\alpha/2}\right)^2 s^2}{E^2} = \frac{(1.645)^2 (28.97)^2}{3^2} \approx 252$$

Since *n* is large enough to make the use of a *z*-value appropriate, no adjustments need be made.

7.40   **a.** The data show that $n = 31$, $\bar{y} = 1.13$, and $s = 0.16$. To calculate a 95% confidence interval, we need the *t* table value (Appendix Table 4) for $31-1= 30$ df and $a = \alpha/2 = 0.05/2 = 0.025$. This value is $t_{0.025,\ 30\ df} = 2.042$

The confidence interval is

$$1.13 - 2.042\left(\frac{0.16}{\sqrt{31}}\right) \le \mu \le 1.13 + 2.042\left(\frac{0.16}{\sqrt{31}}\right)$$

or

$$1.071 \le \mu \le 1.189$$

b. If σ is known to be 0.16, we may use the z table value $z_{0.025} = 1.96$ (rather than the t table value) in the 95% confidence interval.

$$1.13 - 1.96\left(\frac{0.16}{\sqrt{31}}\right) \le \mu \le 1.13 + 1.96\left(\frac{0.16}{\sqrt{31}}\right)$$

or

$$1.074 \le \mu \le 1.186$$

c. The two intervals are almost the same. The artificial (and probably incorrect) assumption that σ is known to equal s allows us to quote a slightly narrower interval.

7.41 With a sample size of only 31 and badly skewed data, we wouldn't put lots of faith in the Central Limit Theorem effect. It does help that the confidence level is a two-sided probability. If the probability in one tail is too small, the probability in the other is probably too large. All in all, the 95% figure is slightly suspect.

Supplementary Exercises

7.42 This is a straightforward exercise in using the confidence interval idea. The exercise indicates that $n = 22$, $\bar{y} = 5.2$, and $s = 3.3$. The exercise indicates that the standard deviation is based on the sample data. Therefore we should use t tables. Appendix Table 4 shows that the t table value for df $= 22 - 1 = 21$ and $\alpha/2 = (1 - 0.90)/2 = 0.05$. The 90% confidence interval is

$$5.2 - 1.721\left(\frac{3.3}{\sqrt{22}}\right) \le \mu \le 5.2 + 1.721\left(\frac{3.3}{\sqrt{22}}\right)$$

or

$$3.99 \le \mu \le 6.41$$

7.43   **a.** No; it's quite unlikely that the data are normally distributed. It is much more plausible that the data would be severely right skewed. Such skewness would make the Empirical Rule work poorly, as seems to be happening. There will be many businesses with small positive margins and a few with very large margins.

b. Even though the confidence interval is two-sided, the small sample size and the probable right skewness of the population indicate that the nominal 90% confidence may be erroneous. A confidence interval assumes that errors above and below the estimate will be symmetric. The 90% figure is dubious.

7.55   Again, we need to calculate a confidence interval for a proportion. We have $n = 500$, which certainly justifies the use of z tables. For $\alpha/2 = 0.025$, the required z table value is 1.96. Recall that the (estimated, large-sample) standard error is $\sqrt{[\hat{\pi}(1-\hat{\pi})]/n}$. The 95% intervals are

(others don't report)

$$0.56 \pm 1.96 \sqrt{\frac{0.56(0.44)}{500}}$$

or

$$0.516 \le \pi \le 0.604$$

(government careless)

$$0.50 \pm 1.96 \sqrt{\frac{0.50(0.50)}{500}}$$

or

$$0.456 \le \pi \le 0.544$$

(overlook it)

$$0.46 \pm 1.96 \sqrt{\frac{0.46(0.54)}{500}}$$

or

$$0.416 \le \pi \le 0.504$$

7.56   If we can assume that the sample was (something close to) a random sample, the editorial stand is nonsensical. The confidence interval calculation allows for random sampling error such as the (tiny) probability of obtaining a sample of the "cheatingest" people. Implicitly, the writer is thinking that the sample is a very small fraction of taxpayers; as we're seen, the issue of the sample fraction is nearly irrelevant to the probable accuracy of an estimate. A more reasonable concern would be whether the data are biased, either in the selection of people to interview or by inducing a particular kind of response.

7.59   **a.** The stem-and-leaf part of the output is the place to look to assess the shape of the data. The data are roughly normally distributed but have a strong peak in the middle. There's no flagrant skewness nor any obvious outliers. We'd call the data quite close to normally distributed.

b. Remember that, for normal data, the sample mean is most efficient, so a mean-based confidence interval should (usually) be narrower than a confidence interval using a robust method such as a median. Note that the confidence interval for the mean is shown as $59.903 \leq \mu \leq 63.017$.

7.60   Recall that normally distributed data plot as a straight line in a normal plot. The data seem to fall quite close to the reference line in the plot. There are a few wiggles at the ends, but nothing substantial. Such a straight line normal plot indicates normal-ish data.

7.66   **a.** The confidence interval is shown under the labels LO 99% CI and UP 99% CI. We might round it off to a couple of decimal places:

$$1.89 \leq \mu \leq 3.94$$

b. The manufacturer's goal of 5 defects per thousand feet equals a mean of 2.5 per 500-foot section. This value is within the interval and is one of the many plausible values. We don't have evidence that the goal is violated, but it might be.

7.67   The systematic tendency (bias) would be to undercount defects. Therefore the mean that was obtained would tend to be too low. The confidence interval should be too low.

7.68   **a.** A negative number of defects is impossible, so the Empirical Rule can't work well for these data. The data presumably aren't mound shaped. The data should be nonnormal, probably skewed. Conceivably an outlier could be inflating the standard deviation.

b. This stem-and-leaf display did not confirm our guess. There is no blatantly obvious skewness, and nothing close to an outlier. The data are perhaps slightly skewed and definitely discrete, taking on only small integer values.

7.69   **a.** Again, simply look at the output. The 95% confidence interval is shown under "`tinterval`" as

$$0.99371 \le \mu \le 1.00597$$

b. This is a misinterpretation of confidence intervals. The confidence interval refers to the mean, not to individual data values. The 95% value refers to the probability that the estimation process will catch the mean within the interval. The interval gives an allowance for error in estimating the mean, not in predicting individual values.

7.70   The boxplot shows one modest outlier on the low side. Otherwise there is no evident skewness. (If you claim it's skewed, tell us which direction!) There's no indication of nonnormal data.

7.75   **a.** Again, we are dealing with a proportion, rather than a mean. If we make no assumption about the value of the true proportion π, we must take the worst case, $\pi = 0.50$. The desired width is $E = 0.04/2 = 0.02$. The required table value is $z_{\alpha/2} = 1.645$ for 90% confidence. Therefore, the required sample size is

$$n = (1.645)^2(0.50)\frac{(1-0.50)}{(0.02)^2} = 1691.27 \text{, rounded up to 1692.}$$

b. If we assume that π is between 0.05 and 0.15, the worst case is the one closest to $\pi = 0.50$, namely $\pi = 0.15$. No other changes in the calculation need to be made so the required sample size is

$$n = (1.645)^2(0.15)\frac{(1-0.15)}{(0.02)^2} = 862.55 \text{, rounded up to 863.}$$

c. The required n in part **b.** is only about half the required sample size in part **a.**, which we'd certainly say was a substantial reduction.

7.76   The sample yields $\hat{\pi}_{\text{not working}} = 61/640 = 0.09531$, $\hat{\pi}_{\text{bypass}} = 28/640 = 0.04375$, and $\hat{\pi}_{\text{failure}} = 33/640 = 0.05156$. For 90% confidence, the required normal table value is $z_{\alpha/2} = 1.645$. The desired confidence intervals are

$$0.04375 - 1.645\sqrt{\frac{(0.04375)(0.95625)}{640}} \leq \pi_{\text{bypass}} \leq 0.04375 + 1.645\sqrt{\frac{(0.04375)(0.95625)}{640}}$$

$$0.05156 - 1.645\sqrt{\frac{(0.05156)(0.94844)}{640}} \leq \pi_{\text{failure}} \leq 0.05156 + 1.645\sqrt{\frac{(0.05156)(0.94844)}{640}}$$

$$0.09531 - 1.645\sqrt{\frac{(0.09531)(0.90469)}{640}} \leq \pi_{\text{not working}} \leq 0.09531 - 1.645\sqrt{\frac{(0.09531)(0.90469)}{640}}$$

or

$$0.030 \leq \pi_{\text{bypass}} \leq 0.057$$
$$0.037 \leq \pi_{\text{failure}} \leq 0.066$$
$$0.076 \leq \pi_{\text{not working}} \leq 0.114$$

7.77 This isn't a simple random sample, because all combinations of meters do not have the same probability of being sampled. Meters in the same sector either are not sampled or are all sampled together. In fact, this sampling method is an example of what is called cluster sampling. Such sampling is useful in situations like this, where taking a simple random sample would require much too much travel time.

7.80 **a.** Again, we are dealing with a proportion, not a mean. The required normal table value is $z_{\alpha/2} = 1.645$ for 90% confidence. The worst-case estimate for π, the proportion of "yes" answers, is 0.50. The desired ± is 0.05. Therefore the required sample size is

$$n = (1.645)^2(0.50)\frac{(1 - 0.50)}{(0.05)^2} = 270.60 \text{, rounded up to 271.}$$

b. Doubling the sample size does not cut the width in half, because the standard error of the sample proportion involves the square root of the sample size. To cut the width in half, we must quadruple the sample size.

7.81 **a.** We are assuming that only half the sampled customers would recommend the dealer. That's an awfully low percentage. We hope that it's not realistic for most dealers!

b. If the proportion is somewhere between 0.80 and 0.95 (so that the worst case, closest to 0.50, is 0.80), the required sample size would be

$$n = (1.645)^2(0.80)\frac{(0.20)}{(0.05)^2} = 173.18 \text{, rounded up to 174.}$$

This is a substantial change.

7.82   **a.** The results you get will depend slightly on which package you use. Here, for example, is Execustat output for the problem.

```
                    Summary Statistics for EX0779

                             TotalExp

        Sample size                  215
        Mean                         5.52372
        Std. deviation               2.59823
```

b. The reported standard deviation is based on the sample, so a *t* interval should be used. For example, Execustat output is as follows.

```
    95% confidence intervals
        Mean: (5.17444,5.873)
        Variance: (5.63421,8.23762)
        Std. deviation: (2.37365,2.87013)
```

The 95% confidence interval is $5.174 \le \mu \le 5.873$.

c. The interval indicates that the true mean expenditure (not individual expenditures) is between 5.174 and 5.873. There is some possibility for error; in the long run, 95% of the intervals constructed in this way include the actual value of the mean.

7.83   **a.** For example, Execustat yielded the following plots.

```
Stem-and-leaf display for TotalExp: unit = 0.1       1|2 represents 1.2

        26      2|2222233555666666777777888
        65      3|000111222223333444455566666677777779999
       (47)     4|00000000000222222222222444445555555577777777999
       103      5|000000222222244455555557777799999
        69      6|000002222222477889
        51      7|0000033455588899
        35      8|03344568
        27      9|000014446
        18     10|00345558
        10     11|0135

               HI|12.2,12.4,12.4,12.8,15.5,15.5
```

The data are right-skewed, with outliers.

b. Certainly, the data are seriously right-skewed, strongly indicating that the underlying population is skewed. However, the Central Limit Theorem effect for a sample size of 215, should be more than ample to make the claimed confidence level correct.

Chapter 8

Hypothesis Testing

8.1   A Test for a Mean, Known Standard Deviation

8.2   Type II Error, β Probability, and Power of a Test

8.1   **a.** Any hypothesis test is concerned with a population parameter, not merely the sample data. The parameter to be tested in this case is the mean waiting time of the entire population of non-emergency patients.

b. In formulating hypotheses, the key issue is whether the research hypothesis should be one-sided or two-sided. In this case, the manager is concerned with mean waiting times that potentially exceed the target value. Therefore, we would take the research hypothesis as one-sided. By taking the research hypothesis as "mean greater than 30," we would require that the data indicate, beyond reasonable doubt, that the target was not being met, before we took action. We would prefer to do so, rather than have the manager overreact to random variation. Therefore,

$$H_0: \ \mu \le 30$$
$$H_a: \ \mu > 30$$

We could take the null hypothesis as $\mu = 30$, because only the boundary value matters in carrying out the test.

c. Assuming no serious problems with the data, we would base the test on the mean for the sample. It is convenient to convert the mean to a z score. Let $\bar{y}$ be the mean waiting time of the sample of 22 patients.

$$\text{Test statistic:} \ \ z = \frac{\bar{y} - \mu_0}{\left(\frac{\sigma}{\sqrt{n}}\right)} = \frac{\bar{y} - 30}{\left(\frac{10}{\sqrt{22}}\right)}$$

Rejection region: At $\alpha = 0.05$, reject H_0 if $z > 1.645$. Note that this z value corresponds to a one-sided hypothesis test—the manager is only concerned that the mean waiting time is too *long*. A picture of the rejection region would have the entire $\alpha = 0.05$ in the upper tail.

8.2   Refer to Exercise **8.1**. We stated the rejection region as $z > 1.645$. We need to calculate the z statistic. Here $\bar{y} = 38.1$. The test statistic, therefore, is,

$$z = \frac{\bar{y} - 30}{\left(\frac{10}{\sqrt{22}}\right)} = \frac{38.1 - 30}{\left(\frac{10}{\sqrt{22}}\right)} = 3.7992$$

Conclusion: Since $z = 3.7992$ falls in the rejection region (it is greater than 1.645), the null hypothesis can be rejected at the 0.05 level.

8.3   We are asked to calculate the probability that the null hypothesis will *not* be rejected, assuming a true (population) mean of 34 minutes. That is, find P(retain $H_0 | H_a$: $\mu = 34$). This is the definition of a β probability.
     If H_0 is rejected whenever

$$z = \frac{\bar{y} - 30}{\left(\frac{10}{\sqrt{22}}\right)} > 1.645$$

then the null hypothesis is rejected whenever

$$\bar{y} > 30 + 1.645\left(\frac{10}{\sqrt{22}}\right)$$

or $\bar{y} > 33.5072$. Therefore,

$$\beta = P(H_0 \text{ not rejected} | H_a \text{ true}) = P(\bar{y} < 33.5072 | \mu = 34)$$

$$= P\left(z < \frac{33.5072 - 34}{\frac{10}{\sqrt{22}}}\right) = P(z < -0.2311) = 0.4086$$

Notice that this assumption does not mean that the sample mean will come out 34 minutes also; the sample mean will vary randomly. We can't just use 34 in the z statistic because we don't know that the *sample* mean will equal 34.
     β probabilities are calculated for other values of μ and summarized in a table below:

$$H_a: \quad \mu = 32 \quad P\left(z < \frac{33.5072 - 32}{\frac{10}{\sqrt{22}}}\right) = P(z < 0.7069) = 0.7602$$

$$H_a: \quad \mu = 36 \quad P\left(z < \frac{33.5072 - 36}{\frac{10}{\sqrt{22}}}\right) = P(z < -1.1692) = 0.1212$$

$$H_a: \quad \mu = 38 \quad P\left(z < \frac{33.5072 - 38}{\frac{10}{\sqrt{22}}}\right) = P(z < -2.1073) = 0.0175$$

$$H_a: \quad \mu = 40 \quad P\left(z < \frac{33.5072 - 40}{\frac{10}{\sqrt{22}}}\right) = P(z < -3.0454) = 0.0012$$

Summary:

μ_a	β
32	0.7602
34	0.4086
36	0.1212
38	0.0175
40	0.0012

Below is a sketch of the β probabilities:

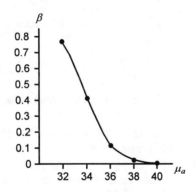

8.4   Waiting times on a busy day would all be long; waiting times on a quiet day would all be short. Hence, independence would be a bad assumption. This is such a crucial assumption that the test would be seriously wrong.

8.3 The *p*-Value for a Hypothesis Test

8.10 **a.** The main question in formulating hypotheses is whether to take a one-sided or two-sided research hypothesis. The exercise indicates that the main concern is that the pack is not working for as long as it should. This fact indicates that we want a one-sided (less than) research hypothesis. The null, "no problem" hypothesis would be that the packs averaged up to standard, 20,000. The appropriate null and research hypotheses are

$$H_0: \quad \mu \geq 20,000$$
$$H_a: \quad \mu < 20,000$$

We could state the null hypothesis as $\mu = 20,000$; only the boundary value is crucial.

b. The test is based on the sample mean, which we normally convert to a *z* statistic. The *p*-value should be one-tailed for a one-sided research hypothesis. The appropriate test statistic and *p*-value are

$$\text{T.S.:} \quad z = \frac{\bar{y} - \mu}{\frac{s}{\sqrt{n}}} = \frac{19,695 - 20,000}{\frac{1.103}{\sqrt{114}}} = -2.9524$$

$$p\text{-value} = P(Z < -2.95) = 0.5 - 0.4984 = 0.0016$$

The table entry for 2.95 is 0.4984; we want the tail area for the *p*-value.

8.11 Since the *p*-value is less than α for the usual $\alpha = 0.1, 0.05, 0.01$, the result is statistically significant at the usual α values. The result is not what we would call "practically significant." If the battery needs recharging after 19,695 calculations instead of 20,000, we seriously doubt that customers will return calculators for this reason (19,695 is about 98.5% of 20,000—this is not serious misadvertising on the part of the manufacturer.)

8.4   Hypothesis Testing with the *t* Distribution

8.12   The exercise states that we want a one-sided research hypothesis. The standard deviation is based on the sample data, not on the underlying population. Therefore, we want a one-tailed test using a *t* statistic.

A sample of 18 two week periods yielded a mean of 1,718.3 ($\bar{y} = 1{,}718.3$) and a standard deviation of 137.8 ($s = 137.8$). The sample information is used to test the research hypothesis $\mu > 1{,}600$.

H_0: $\mu = 1{,}600$

H_a: $\mu > 1{,}600$

T.S.: $t = \dfrac{\bar{y} - \mu_0}{\frac{s}{\sqrt{n}}} = \dfrac{1{,}718.3 - 1{,}600}{\frac{137.8}{\sqrt{18}}} = 3.64$

R.R.: At $\alpha = 0.1$, reject H_0 if $t > 1.333$; this value is found in Appendix Table 4

with 17df in the 0.10 column.

Conclusion: Since the calculated *t*-value lies well into the rejection region, reject H_0

at the 0.1 level.

8.13   Because the *t* test rejected the null hypothesis at $\alpha = 0.10$, it follows by the Universal Rejection Region that the *p*-value is less than 0.10. We can say more by trying other table values.

$$p\text{-value} = P(t > t_{\text{actual}})$$

Using the *t* table with 17 df, since

$$P(t_{df=17} > 2.898) = 0.005$$

then

$$P(t_{df=17} > 3.64) < 0.005$$

so the *p*-value < 0.005

The actual t statistic of 3.64 is just barely within the tabled value for $\alpha = 0.001$, namely 3.646. Therefore, we can't quite say that the p-value is less than 0.001.

Yes, we would say that the research hypothesis is strongly supported since the p-value is *very* small.

8.17 **a.** The output shows means and standard deviations. We must look for the information about the differences.

$$\bar{d} = 1.8$$
$$s_d = 1.988858$$

b. First, notice that the average difference (Postal – Private) is positive, suggesting that the private shipper has a shorter average. The test checks to see whether the apparent difference could reasonably have arisen by luck. To calculate a one-tailed p-value, divide the two-tailed value by 2 and obtain 0.00935. This p-value is (barely) less than $\alpha = 0.01$, so we can support the research hypothesis.

8.5   Assumptions for t Tests
8.6   Testing a Proportion: Normal Approximation

8.18 **a.** If customers are equally divided, half will prefer the airbag. Denoting the proportion in the whole population (not just our sample) preferring the airbag as π, the null hypothesis is

$$H_0: \quad \pi = 0.50$$

b. We don't have a particular direction in mind. We would like to find out if a majority of customers have a preference in either direction. Therefore, we would use a two-sided research hypothesis. In the output, there is a line saying "Test of mu - 0.5000 vs mu not = 0.5000," which indicates that a two-sided research hypothesis was, in fact, used.

c. We have a choice of tests here. Technically, the z test exactly matches the test described in the text, so we'll use that one. We could formulate a rejection region and compare the computed z statistic to a table value. For a two-tailed test with $\alpha = 0.01$, the rejection region is $|z| > 2.58$, using Appendix Table 3. The actual value of the statistic is shown as 2.71. Therefore we reject the null hypothesis. Another way to say that is to say that there is a statistically significant difference from $\pi = 0.5$. Another, quicker way to do the same thing is to use the indicated p-value. It is shown in the output as 0.0068, which is smaller than $\alpha = 0.01$. Once again, we find a statistically significant result. In practice, more than half our population of customers prefers the airbag to the discount.

It makes almost no difference whether we consider the z test or the t test in the output, because the sample size and therefore the df are so large. The p-values are practically identical.

8.19 The p-value is an index of how good our evidence is that something is happening. It is not an indicator of how strong that "something" is. In this case, it appears that 53.4% of our customers prefer the airbag. To us, that is not a huge deviation from 50%. It is big enough that we're reasonably sure it's not random, but it does not qualify as "dramatic," as far as we're concerned.

8.20 The text indicates that the guideline for the adequacy of this normal approximation is the expected number of successes and the expected number of failures. Both should be larger than 5 or so. Assuming that the null hypothesis is true, as we did in computing the p-value, the expected numbers of successes and failures should each be half the sample size, namely 793. This is far larger than the guidelines require. The approximation is fine.

8.7   Hypothesis Tests and Confidence Intervals

8.26 **a.** The output in the exercise shows a 95% confidence interval for the mean, corresponding to $\alpha = 0.05$. The interval is $4.4152 \leq \mu \leq 5.9398$. This interval certainly includes $\mu = 4.62$. Therefore, we must retain H_0: $\mu = 4.62$; this is one plausible value for the mean in this city.

b. The staff member is correct in saying that the deviation was not statistically significant, because that simply means that the null hypothesis can't be rejected. But the interpretation is not valid. The null hypothesis must be retained, but isn't proved. The confidence interval also includes a wide range of other plausible values for μ, ranging from considerably lower than 4.62 to substantially higher. There is still quite a bit of uncertainty about the correct value for μ.

8.27 The skewness refers to individual claims data. We know from the Central Limit Theorem that the theoretical distribution of the sample mean is nearly normal for large enough n. In this case, n is 187, large enough to be confident that normal probabilities are quite accurate for the sample mean, even though the population of individual values may be quite seriously skewed.

8.31   **a.** For such a small sample size, we clearly must use t tables. There are $n-1=4$ df, and we want a 95% confidence interval. The required table value is $t_{\alpha/2} = t_{0.025,\ 4\ \text{df}} = 2.776$. The confidence interval is

$$11{,}880.4 - 2.776\left(\frac{798.68}{\sqrt{5}}\right) \le \mu \le 11{,}880.4 + 2.776\left(\frac{798.68}{\sqrt{5}}\right)$$

or

$$10{,}888.9 \le \mu \le 12{,}8719$$

b. The value $\mu = 11{,}260$ is well within the 95% confidence interval, corresponding to $\alpha = 0.05$. Therefore, it must be retained as one plausible value.

8.32   No, it isn't. The confidence interval is extremely wide, because of the large amount of variability and the small sample size. It is possible that the mean volume has not increased. It is also possible that it has increased by as much as 1,600 vehicles (which would be about a 15% increase) or decreased by as much as 300 vehicles. The confidence interval is so wide that it isn't possible to say much.

8.33   Notice that the volumes clearly are increasing. Except for a very slight dip on the fourth day, the volumes increased each day. There appears to be an upward trend in the data. This is a plausible result, in that traffic flow should improve as motorists become more accustomed to the new arrangements. Therefore, the average value in the sample probably underestimates the long-run average volume.

Supplementary Exercises

8.39   The following arguments, among others, should apply.
   One sided alternative: Consumers will benefit if the mpg is underestimated; therefore, we should only be concerned if the mpg is overestimated and the consumer is misled by misadvertising. Thus we should only be concerned whether the ratings are too high.
   Two sided alternative: For technological reasons, a two sided alternative may be desirable. For example, if the EPA set minimum standards to be attained, at say 30 mpg, manufacturers may want to know how near they are to obtaining the standard. Also manufacturers may be interested in the effect of pollution control devices on mpg and a two sided alternative may be appropriate (are they over- or underestimating effects?).

8.40 The sample of 8 cars yielded a sample mean gas mileage of 26.7 ($\bar{y} = 26.7$).
Assume the population standard deviation is 2.1 ($\sigma = 2.1$). We could calculate a 99%
($\alpha = 0.01$) confidence interval and see if the null hypothesis value, 28.2, is included.
We could figure out a p-value and compare it to $\alpha = 0.01$. For illustration, we'll carry
out a formal five-step test.

H_0: $\qquad \mu = 28.2$

H_a: $\qquad \mu \neq 28.2$ (specified as two-sided in the exercise)

T.S.: $\qquad t = \dfrac{\bar{y} - \mu_0}{\frac{\sigma}{\sqrt{n}}} = \dfrac{26.7 - 28.2}{\frac{2.1}{\sqrt{8}}} = -2.0203$

R.R.: $\qquad$ At $\alpha = 0.1$, reject H_0 if $z < -2.58$ or $z > 2.58$

Conclusion: Since z does not lie in the region, do not reject H_0.

8.41 Remember that the p-value is the probability of a value at least as extreme as the
actually obtained test statistic. Our actual z statistic came out 2.02. The entry in
Appendix Table 3 for 2.02 is 0.4783, so the one-tail area beyond 2.02 is
$0.5 - 0.4783 = 0.0217$. Double that for the two-tail area, because we have a two-sided
research hypothesis.

$$p\text{-value} = 2P(Z \geq |z_{actual}|) = 2P(Z \geq 2.02) = 0.0434$$

8.42 The official is making the error of thinking that retaining a null hypothesis proves that it
is true. We don't have enough evidence to say that the claim is wrong. With a
sample of size 8, we don't have much evidence, period. We certainly shouldn't say
that the mean is exactly 28.2000000 mpg. Furthermore, although the result was not
significant at the 0.01 level, it *is* significant at the 0.05 and 0.1 levels (because the p-
value is 0.0434). We shouldn't accept 28.2 mpg as absolute truth.

8.43 The standard z (population standard deviation assumed known) confidence interval
for μ is

$$\bar{y} - z_{\alpha/2} \times \frac{\sigma}{\sqrt{n}} \leq \mu \leq \bar{y} + z_{\alpha/2} \times \frac{\sigma}{\sqrt{n}}$$

where in our case

$$\alpha = 0.01; \ z_{\alpha/2} = z_{0.005} = 2.576 \text{ (or 2.58 if you prefer)}$$

Therefore,

$$26.7 - 2.576 \times \frac{2.1}{\sqrt{8}} \le \mu \le 26.7 + 2.576 \times \frac{2.1}{\sqrt{8}}$$

$$26.7 - 1.9126 \le \mu \le 26.7 + 1.9126$$

$$24.7874 \le \mu \le 28.6126$$

rounding

$$24.8 \le \mu \le 28.6$$

We can "reliably assume" with 99% confidence that the true mean mpg lies in the interval [24.8, 28.6].

8.47   **a.** The mean referred to is for the entire population of calls. Therefore it should be denoted μ. The department's research hypothesis is H_a: $\mu < 20$. The null hypothesis H_0 is the opposite of H_a. In principle, we should take H_0: $\mu < 20$. However, only the boundary case in the null hypothesis needs to be considered, so we take H_0: $\mu = 20$.

b. Because we are assuming a "normalish" population and a known population standard deviation, we can use a z statistic:

$$z = \frac{\bar{y} - \mu_0}{\frac{\sigma}{\sqrt{n}}}$$

c. We have already defined H_0 and H_a, as well as the test statistic z. The rejection region is determined by the requirement that the one-tailed α probability should be 0.05. We reject H_0 if $z < -1.645$. (See the figure.) The actual value of z is shown in the output as $z = 1.52$. Therefore, we can't reject H_0 and must retain it.

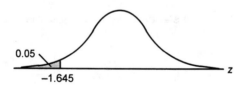

8.48   The p-value is the probability that a test statistic will be at least as extreme as the observed value. In this situation, the observed value is $z = 1.52$ and we want to consider only smaller (more negative) values as being "at least as extreme." In other words, we want a one-tailed p-value. The output shows this value as 0.065.

8.49   The data are clearly right skewed, strongly indicating that the entire underlying population is likewise skewed. The sample size is relatively small, $n = 29$. The Central Limit Theorem will not work fully in this case. (Also, we have a one-tailed situation, so that we won't have one too-large probability compensating for another, too-small probability.) The claimed α probability may well be incorrect.

8.50 Because we are no longer assuming that α is known, we must use a t statistic rather than a z statistic. The indicated t table value with $n-1=28$ df and one-tail area 0.05 is 1.701; for a left-tail test, reject H_0 if $t < -1.701$. The actual value of the test statistic is shown as $t = -1.45$. We can't reject H_0 and must retain it. If you prefer, another way to carry out the test is to note that the p-value is 0.080, which is greater than 0.05. Once again, we retain the null hypothesis.

8.55 **a.** Notice that now we are dealing with a proportion, rather than a mean. Denote the long-run proportion of wins as π. If the firm really will win half the time, $\pi = 0.50$. This is our null hypothesis (no difference between the firm and its chief competitor).

b. The sample proportion is $\hat{\pi} = 5/16 = 0.3125$. Use the approximate z test in the text.

$$z = \frac{\hat{\pi} - \pi_0}{\sqrt{(\pi_0)\frac{(1-\pi_0)}{n}}} = \frac{0.3125 - 0.5}{\sqrt{(0.5)\frac{(1-0.5)}{16}}} = -1.50$$

The z table value corresponding to a one-tail area of 0.05 is 1.645. Therefore, we would reject H_0 if $z < -1.645$. That is not the case, so we can't reject the null hypothesis; can't support the research hypothesis.

c. This misinterpretation of hypothesis tests says that retaining the null hypothesis proves it's true. Not so; all we can say is that we don't have enough evidence to say that that hypothesis is false. With a sample size of only 16, and with each observation providing very little information (merely a win or a loss), we don't have much evidence at all. The president simply can't say yet whether the v.p. will win half the time in the long run.

8.56 The guidelines for using the normal approximation say that the expected numbers of successes and of failures should be at least 5. Assuming the null hypothesis, as we did in computing our probabilities, the expected numbers of wins and of losses should be $16(0.50) = 8$. Therefore, our guidelines are met, though not by much. The approximation is at least decent. Note that that does not imply that our test is very effective. With such a small n, it will not have much power.

8.67 **a.** Here are results from Excel, which should be virtually identical to those given by other packages.

```
                LeaveDay

Mean                         5.04
Standard Error               0.569
Median                          4
Mode                            1
Standard Deviation           4.02
```

The mean is 5.04, the median is 4.00, and the standard deviation is 4.02. As always, your computer program may report answers to a different number of decimal places. The fact that the mean is a full quarter of a standard deviation larger than the median suggests that the data are severely right skewed. Furthermore, negative values are logically impossible, but the mean minus two standard deviations is –3. Again it seems likely that the data are right skewed.

b. A histogram obtained from Excel is shown here. A histogram from another program might differ in detail, but will show the same overall pattern. The data are evidently right skewed.

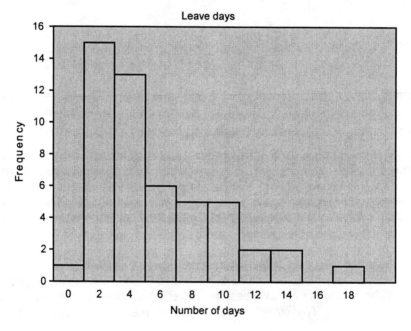

8.68 **a.** Here is a backhanded way of testing the mean using Excel. The "gimmick" is to enter a whole column of values equal to the null hypothesis mean, 5.7, and to use a "paired sample *t* test," which will be discussed in the next chapter. Admittedly, this is a trick. Other packages should allow us to do the same thing more naturally.

```
t-Test: Paired Two Sample for Means

                             LeaveDay NullHyp
Mean                             5.04     5.7
Variance                    16.16163
Observations                      50      50
Hypothesized Mean Differen         0
df                                49
t Stat                        -1.161
P(T<=t) one-tail               0.126
t Critical one-tail            1.677
P(T<=t) two-tail               0.251
t Critical two-tail            2.010
```

The t statistic comes out −1.16, with 49 degrees of freedom. For any conventional α value, the null hypothesis of no change must be retained.

b. The two-tailed p-value is about 0.251, and the one-tailed value is about 0.126. Both are shown in the Excel output. Different packages will report different numbers of decimal places. The one-tailed value seems more appropriate in this situation. Because the manager is trying to reduce the number of leave days, we should take $H_a: \mu < 5.7$. In any case there is not enough evidence to say that the policy has done what was intended.

Review Exercises—Chapters 7–8

R.45 **a.** Assuming that σ is known to be 35.4, we may base the confidence interval on the z table (Appendix Table 3). The required area for a 99% confidence interval is 0.495; note that if the area under the curve between 0 and z is 0.495, the right-tail area is 0.005, as is the left tail area. The two-tail area is $2(0.005) = 0.01$, as desired. Looking through Appendix Table 3, we find that $z = 2.57$ corresponds to area 0.4949 and $z = 2.58$ corresponds to 0.4951. We could either use 2.575 or be conservative and use 2.58.

The form of the interval is

$$\bar{y} - z_{\alpha/2}\frac{\sigma}{\sqrt{n}} \le \mu \le \bar{y} + z_{\alpha/2}\frac{\sigma}{\sqrt{n}}$$

Using $z_{0.01/2} = 2.58$ we have

$$253.32 - 2.58\left(\frac{35.4}{\sqrt{50}}\right) \le \mu \le 253.32 + 2.58\left(\frac{35.4}{\sqrt{50}}\right)$$

$$240.40 \le \mu \le 266.24$$

(had we used $z_{\alpha/2} = 2.575$, we would have gotten $240.43 \le \mu \le 266.21$.)

The output shows the result as $240.42 \le \mu \le 266.22$. For all practical purposes, we have the same answer by hand or by computer.

b. The sample mean is inefficient if the population is distinctly nonnormal. The shape of the sample data will suggest what the population shape is. In the normal plot, the data plot very close to a line. The sample data appear quite normally distributed. The sample mean should be a reasonably efficient estimator of the population mean.

R.46 We need only check the confidence interval. We assume a two-sided research hypothesis corresponding to the ± form of the confidence interval. The $\alpha = 0.01$ value corresponds to a $100(1-0.01)\% = 99\%$ confidence interval. The confidence interval does not include $\mu = 230.2$. Therefore $H_0:\ \mu = 230.2$ is rejected at $\alpha = 0.01$.

R.47 The null hypothesis is $H_0:\ \mu = 230.2$, and the research hypothesis is $H_a:\ \mu \neq 230.2$. The test statistic (assuming that σ is known to be 35.4) is

$$\text{T.S.:}\ Z = \frac{\bar{y} - 230.2}{\frac{35.4}{\sqrt{50}}}$$

The rejection region (for $\alpha = 0.01$, two-tailed) is $|Z| > 2.58$ or perhaps $|Z| > 2.575$. The actual value is

$$Z = \frac{253.32 - 230.2}{\frac{35.4}{\sqrt{50}}} = 4.62$$

Therefore, we reject H_0 most emphatically.

R.48 Because the test in Exercise **R.47** was two-sided, the p-value should also be two-sided (often called two-tailed).

$$p\text{-value} = 2P(|Z| > 4.62)$$

There is no entry for $z = 4.62$ in Appendix Table 3. Rather than trying to be fancy with interpolation, we use the closest smaller value in the table, $z = 4.50$. The area under the normal curve between 0 and 4.50 is shown as 0.49999660. Thus

$$p\text{-value} \approx 2P(Z > 4.50) = 2(0.5 - 0.49999660) = 0.0000068$$

The output shows the p-value as 0.0000 to four decimal places. We obtained the same thing.

R.49 When the population standard deviation σ is not assumed to be known, we must use the t table (with $50 - 1 = 49$ df) rather than the z table. For a 99% confidence interval, we need the table value $t_{0.005, \, 49 \, df}$. Appendix Table 4 does not have entries for 49 df. The closest entries are for 40 df and for 60 df; $t_{0.005, \, 40 \, df} = 2.704$ and $t_{0.005, \, 60 \, df} = 2.660$. We could be conservative and use the 40 df value, or we could interpolate.

$$t_{0.005, \, 49 \, df} \approx \frac{11}{20}(2.704) + \frac{9}{20}(2.660) = 2.684$$

Using the interpolated value and the sample standard deviation shown in the output, the 99% confidence interval is

$$\bar{y} - t_{0.005} \frac{s}{\sqrt{n}} \le \mu \le \bar{y} + t_{0.005} \frac{s}{\sqrt{n}}$$
$$253.32 - 2.684 \left(\frac{36.1}{\sqrt{50}} \right) \le \mu \le 253.32 + 2.684 \left(\frac{36.1}{\sqrt{50}} \right)$$
$$239.62 \le \mu \le 267.02$$

The Minitab output shows virtually the same answer, $239.64 \le \mu \le 267.00$. This interval doesn't include $\mu = 230.2$, so we reject H_0: $\mu = 230.2$ at $\alpha = 0.01$. (if we had used $t_{0.005, \, 40 \, df} = 2.704$, we would have gotten the slightly wider interval $239.52 \le \mu \le 267.12$; and again would have rejected H_0: $\mu = 230.2$.)
   A formal test would use

$$t = \frac{\bar{y} - 230.2}{\frac{s}{\sqrt{n}}} = \frac{253.32 - 230.1}{\frac{36.1}{\sqrt{50}}} = 4.55$$

We do not have t table values for 49 df, but $t = 4.55$ is well beyond either $t_{0.001, \, 40 \, df} = 3.307$ or $t_{0.001, \, 60 \, df} = 3.232$. Thus $P(t > 4.55) < 0.001$ and

$$p\text{-value} = 2P(t > 4.55) < 2(0.001) = 0.002$$

Minitab output shows the same value for t, and a p-value of 0.0000. We have effectively the same results. We get the same conclusion using t as using Z. Reject H_0: $\mu = 230.2$ conclusively.

R.50 Again, see if the null hypothesis value is included in the confidence interval. Whichever level of "ACHIEVED CONFIDENCE" we use, the interval does not include 230. Therefore, reject H_0: population median $= 230$.

R.51   The width of the confidence interval for the mean is $267.02 - 239.62 = 27.40$. The width of the confidence interval for the median is $275 - 233 = 42.0$. The narrower confidence interval for the mean indicates that the sample mean is more efficient, confirming the judgment we made that the data were nearly normally distributed.

R.62   Use the approximate z interval from the text. The sample proportion is $\hat{\pi} = 22/133 = 0.1654$. For a 95% confidence interval $z_{\alpha/2} = z_{0.025} = 1.96$. The desired confidence interval is

$$\hat{\pi} - z_{\alpha/2}\sqrt{\frac{\hat{\pi}(1-\hat{\pi})}{n}} \le \pi \le \hat{\pi} + z_{\alpha/2}\sqrt{\frac{\hat{\pi}(1-\hat{\pi})}{n}}$$

$$0.1654 - 1.96\sqrt{\frac{0.1654(0.8346)}{133}} \le \pi \le 0.1654 + 1.96\sqrt{\frac{0.1654(0.8346)}{133}}$$

$$0.102 \le \pi \le 0.229$$

R.63   **a.** Here we must calculate a sample size to yield a specified degree of random error. The desired $E = 0.03$. Assuming $\hat{\pi} = 22/133$, we must take

$$n = \frac{z_{\alpha/2}^2 \hat{\pi}(1-\hat{\pi})}{E^2} = \frac{(1.96)^2\left(\frac{22}{133}\right)\left(\frac{111}{133}\right)}{(0.03)^2} = 589.3$$

Rounding up, we find that the required $n = 590$.

b. Without any assumption about the likely value of $\hat{\pi}$, we take the worst case, $\hat{\pi} = 0.50$. Again, the desired $E = 0.03$.

$$n = \frac{(1.96)^2(0.50)(1-0.50)}{(0.03)^2} = 1{,}067.1$$

which we round up to 1,068.

R.64   The guideline for using a normal approximation to the binomial is that the expected numbers of successes and failures should both be at least 5. We don't know π, the population proportion, so we estimate it by $\hat{\pi} = 22/133$. The estimated expected frequencies are $133(22/133) = 22$ and $133(111/133) = 111$, both considerably larger than 5. In Exercise **R.63**, the sample size is much larger, so the normal approximation should be even better.

R.65   For a random sample, every transaction should have the same probability of being sampled. In this sampling method, large transactions have higher probability of being sampled. At the extreme a $1 transaction has a very low probability of being sampled and a \$99,999 transaction is almost certain to be sampled. There is a size bias favoring large accounts.

R.66   **a.**   The population standard deviation is not known; the specified standard deviation is $s = 2{,}271$. In principle, we should use t methods. The desired table value is

$$t_{\alpha/2,\ n-1\ df} = t_{0.025,\ 240\ df} = 1.970$$

The confidence interval is

$$\bar{y} - t_{\alpha/2}\frac{s}{\sqrt{n}} \le \mu \le \bar{y} + t_{\alpha/2}\frac{s}{\sqrt{n}}$$

$$5{,}381 - 1.970\left(\frac{2{,}271}{\sqrt{241}}\right) \le \mu \le 5{,}381 + 1.970\left(\frac{2{,}271}{\sqrt{241}}\right)$$

$$5{,}093 \le \mu \le 5{,}669$$

(If we had used $z_{0.025} = 1.96$ instead of $t_{0.025,\ 240\ df} = 1.970$, we would have gotten $5{,}094 \le \mu \le 5{,}668$.)

b.   The Central Limit Theorem (CLT), with a sample size $n = 241$ guarantees that normal or t probabilities will be extremely close to correct, even if the population isn't normal. Thus the right-skewness of the data is not the issue. The major problem is that the biased sampling in favor of large accounts will tend to given a considerable overestimate of the population mean.

Chapter 9

Comparing Two Samples

9.1 Comparing the Means of Two Populations

9.1   **a.** The basic issue is: Which part of the output is relevant? We are first assuming equal population variances. Therefore we should use the pooled *t* method. The output under "TwoT" with subcommand "Pooled" contains results for this method. That part of the output shows the confidence interval as

$$-1.26 \leq \mu_1 - \mu_2 \leq 0.56$$

When we do not assume equal variances, the output under "TwoT" without the "Pooled" subcommand gives the *t'* results. Here the confidence interval is

$$-1.28 \leq \mu_1 - \mu_2 \leq 0.58$$

The intervals are virtually identical, so it makes almost no difference which is used.

b. There are several ways to use the output to answer this question. We could look at the confidence intervals and note that a difference of 0 is well within the reasonable range of possibilities; therefore retain that null hypothesis. Alternatively, we could use the calculated *t* statistics, shown as T = -0.80 in both parts of the output. Compare the calculated statistic to the tabled values for tail area 0.025 ($\alpha = 0.05$, two-tailed) and you'll see that we must retain the null hypothesis. Or still another way: Look at the *p*-values (0.43 or 0.44). They aren't even close to being less than α. However you wish to do it, we cannot conclude that there is a nonzero difference.

9.2   The *p*-value is shown as 0.435 for the pooled-variance (equal variance) test and 0.438 for the separate variance (unequal variance) test. The results are very close, and in either case, the result is far from being statistically significant.

9.3   The assumptions we make are: unbiased samples, independence between (and within) samples, normal population distributions, and equal variances in the pooled-variance approach. We have no information about bias. There is no reason to think that there is any relation between the samples, so the independence assumption seems reasonable. The boxplots appear modestly left skewed, but not seriously so. The normality assumption is mildly questionable, at worst. The two boxplots clearly show different amounts of variability, so the equal variance assumption is most suspect. However, the pooled-variance method gives the same results as the separate-variance t' method, so there is no reason to doubt the conclusions.

9.4   **a.** In this simulation, H_0 is true. Both population means are equal to 50. Note that $\sigma_1 > \sigma_2$ and $n_1 > n_2$. In the text, it was stated that the nominal α probabilities for the pooled-variance t test could be seriously in error in this situation. We would expect the t' test to be better, in the sense that the claimed α probabilities would be closer to correct.

b. If the tests are working correctly, the null hypothesis should be erroneously rejected in 50 of the 1000 samples. The actual frequency is much higher (203 of 1000) for the pooled-variance t test, indicating that the actual α probability is much higher than claimed. The actual frequency for the t' test is very close to the nominal 5%. Therefore, we find, as we thought, that the t' test is better in this situation, in the sense that the claimed α probability is much more believable.

9.6   **a.** The pooled standard deviation is shown as POOLED STDEV = 9.38. If we wish to check the calculation by hand, we find

$$s_p^2 = \frac{(n_1 - 1)s_1^2 + (n_2 - 1)s_2^2}{n_1 + n_2 - 2}$$

$$= \frac{(10 - 1)(10.46)^2 (16 - 1)(8.67)^2}{10 + 16 - 2} = 88.010$$

and

$$s_p = \sqrt{88.010} = 9.38$$

b. The confidence interval is shown as $2.9 \le \mu_A - \mu_B \le 24.1$. If we wish to check this by hand, we need a t table value from Appendix Table 4 with $n_1 + n_2 - 2 = 24df$. For a 99% confidence interval we want $a = \alpha/2 = 0.005$. The desired value is $t_{0.005, \, 24df} = 2.797$. The interval is

$$(\bar{y}_A - \bar{y}_B) - t_{\alpha/2} s_p \sqrt{\frac{1}{n_A} + \frac{1}{n_B}} \le \mu_A - \mu_B \le (\bar{y}_A - \bar{y}_B) + t_{\alpha/2} s_p \sqrt{\frac{1}{n_A} + \frac{1}{n_B}}$$

$$(51.50 - 38.00) - 2.797\sqrt{88.010}\sqrt{\frac{1}{10} + \frac{1}{16}} \le \mu_A - \mu_B \le (51.50 - 38.00) + 2.797\sqrt{88.010}\sqrt{\frac{1}{10} + \frac{1}{16}}$$

$$2.92 \le \mu_A - \mu_B \le 24.08$$

c. We have H_0: $\mu_A = \mu_B$ or equivalently H_0: $\mu_A - \mu_B = 0$. The 99% confidence interval (corresponding to $\alpha = 0.01$) calculated in part **b.** doesn't include 0.00. Therefore we reject H_0.

9.7 The confidence interval is shown in the output (in the "TwoT" part without the "Pooled" subcommand) as

$$1.9 \le \mu_A - \mu_B \le 25.1$$

To check this by hand, if desired, we need approximate df for the t' procedure.

$$c = \frac{\frac{s_1^2}{n_1}}{\frac{s_1^2}{n_1} + \frac{s_2^2}{n_2}} = \frac{\frac{(10.46)^2}{10}}{\frac{(16.46)^2}{10} + \frac{(8.67)^2}{16}} = 0.6996$$

$$df = \frac{(n_1 - 1)(n_2 - 1)}{(n_2 - 1)c^2 + (n_1 - 1)(1 - c)^2} = \frac{(10 - 1)(16 - 1)}{(16 - 1)(0.6996)^2 + (10 - 1)(0.3004)^2} = 16.56$$

rounded down to 16. From Appendix Table 4, $t_{0.005,\ 16df} = 2.921$.
The 99% confidence interval is

$$(\bar{y}_A - \bar{y}_B) - t_{\alpha/2} \sqrt{\frac{s_A^2}{n_A} + \frac{s_B^2}{n_B}} \le \mu_A - \mu_B \le (\bar{y}_A - \bar{y}_B) + t_{\alpha/2} \sqrt{\frac{s_A^2}{n_A} + \frac{s_B^2}{n_B}}$$

$$(51.50 - 38.00) - 2.921\sqrt{\frac{(10.42)^2}{10} + \frac{(8.67)^2}{16}} \le \mu_A - \mu_B \le (51.50 - 38.00) + 2.921\sqrt{\frac{(10.42)^2}{10} + \frac{(8.67)^2}{16}}$$

$$1.95 \le \mu_A - \mu_B \le 25.05$$

Once again, 0.00 is not included in the interval, so H_0 is rejected by either method. The confidence interval using t' is wider than the interval using the pooled-variance. However, because $n_1 < n_2$ and $s_1 > s_2$ it is dangerous to assume equal population variances, as we do in using pooled-variance methods. The claimed 99% confidence interval using t' is more believable.

9.2   A Nonparametric Test: The Wilcoxon Rank Sum Test

9.16   **a.** The data are already ordered, so that it's easy to rank them. The ranks for source I are 1, 2, 3, 4, 6, 7, 10, 14, 17, 19. The rank sum is $T_1 = 1 + 2 + \cdots + 19 = 83$. We have $n_1 = n_2 = 10$

$$z = \frac{T_1 - \frac{n_1(n_1 + n_2 + 1)}{2}}{\sqrt{\frac{n_1 n_2(n_1 + n_2 + 1)}{12}}} = -1.66$$

For $\alpha = 0.05$ (two-tailed), reject H_0 if $|z| = z_{0.025} = 1.96$. Because $|z| = 1.66 < 1.96$, H_0 must be retained.

b. The p-value may be calculated directly from the z table. We have a two-sided research hypothesis, so we want a two-tailed p-value.

$$p\text{-value} = 2P(Z > 1.66) = 2(0.5 - 0.4515) = 0.0970$$

9.17   Plots of the data (or simply a look at the numbers) show that both samples are severely right-skewed. Therefore we may assume that the underlying populations are badly right-skewed as well. A t test is most effective for normal populations. Therefore a rank test should work better for these data.

9.3   Paired-Sample Methods
9.4   The Signed Rank Method

9.18   **a.** The paired sample t test is shown in the output under "TTest 0.0 'Diff'" as t = 4.05 with a p-value of 0.0012. We could look up the appropriate t table value, but the easiest way to carry out the test is to note that the p-value is less than $\alpha = 0.10$. We have clear evidence of a real difference. Reject the null hypothesis.

b. The 90% confidence interval for μ_d is shown in the output under "TInterval."

$$1.959 \leq \mu_d \leq 4.974$$

The interval does not include 0.000, so again we support the research hypothesis.

9.19   **a.** The Wilcoxon signed rank test result is shown in the output under "WTest" to be 98.0 The p-value is shown as 0.005. The p-value is less than $\alpha = 0.10$, so we have evidence to reject the null hypothesis.

b. The p-value for the signed rank test is larger than the one for the t test, and therefore somewhat less conclusive. However, the basic conclusion of both this test and the t test indicates that Display A has a better effect on sales than Display B.

c. The stem-and-leaf display is somewhat left skewed. However, it is not so severe that we would doubt the conclusion of either test, especially because the p-values were so far below α. It's not a close call.

9.20   The p-value for the t test is shown as 0.0012; for the "WTest" (signed rank test), it is shown as 0.005.

9.21   **a.** Since $n_1 = n_2$, t will have the same value as t'. Use of the t test will be reasonably accurate (assuming the alternative approach so that the samples are independent). The output shows $t = 0.75$, with a p-value of 0.46. Because the p-value is much larger than α, we cannot come even close to rejecting the null hypothesis. By this test, there would be no evidence to conclude there is a real difference.

b. The variability in sales volumes among the stores completely masks the variability due to different displays when the data are treated as independent samples. The pairing process effectively eliminates this undesirable source of variation. As a result, the tests using pairing are much more conclusive.

9.22   The binomial (sign) test results in the output show a p-value of 0.0129. If we use $\alpha = 0.10$ again, we would reject the null hypothesis conclusively.

H_0 : The median difference is 0. (The proportion of positive differences equals the proportion of negative differences.)

H_a : The median difference is not 0.

T.S.: Let Y = number of differences $> 0 = 12$
(one difference $= 0$, eliminate it, $n = 14$)

R.R.: For $n = 14$, $\pi = 0.5$, reject H_0 if $Y \le 3$ or $Y \ge 11$ for $\alpha = 0.01$.
$P(Y \le 3 \text{ or } Y \ge 11) = 2(0.0287) = 0.0574$

Conclusion: Reject H_0 since $Y = 12$ lies in the rejection region. The data indicate that the median difference is greater than 0.

Supplementary Exercises

9.31   **a.** The t statistic is shown in the output as $T = 0.80$ with 22df. The p-value is given as $P = 0.43$. The p-value is much larger than $\alpha = 0.10$, so we can't come close to supporting the research hypothesis.

b. The output doesn't include any information about a rank test, so we'll need to carry this one out by hand.
The ranks of the observations are given below:

Variety R	37.1	47.3	53.2	54.0	58.7	59.8	61.1	61.6	64.2	68.0	71.1	74.6
Rank	2.0	3.0	8.0	9.0	11.0	12.0	15.0	16.0	19.0	21.0	22.0	24.0

Variety K	32.4	48.6	49.0	49.0	50.8	57.4	59.9	60.0	61.9	62.1	66.7	72.0
Rank	1.0	4.0	5.5	5.5	7.0	10.0	13.0	14.0	17.0	18.0	20.0	23.0

R — sum of ranks = 162
K — sum of ranks = 138

Let T denote the sum of ranks for Variety R = 162

$$\mu_T = \frac{n_1(n_1 + n_2 + 1)}{2} = \frac{12(25)}{2} = 150$$

$$\sigma_T^2 = \frac{n_1 n_2}{12}(n_1 + n_2 + 1) = \frac{12(12)(25)}{12} = 300, \quad \sigma_T = \sqrt{300}$$

The test follows as:

H_0 :   The two populations are identical.

H_a :   Population 1 is shifted to the right of Population 2.

T.S.: $z = \dfrac{T - \mu_T}{\sigma_T} = \dfrac{162 - 150}{\sqrt{300}} = 0.69$

R.R.:   For $\alpha = 0.1$, reject H_0 if $z > 1.282$

Conclusion: Do not reject H_0

9.32   The histograms don't provide much evidence with such small sample sizes. From what we can see there are no obvious outliers, nor is there extreme skewness. The assumption of normality of the populations seems tolerably appropriate here. In such populations, the t test is more powerful. However, since the conclusion of each of the tests is retention of the null hypothesis, choosing the more appropriate test is not critical.

9.33   In the output, the two-tailed p-value for the t test is shown as 0.43. Technically, we should cut it in half, because the research hypothesis is one-sided. So we'd report it as 0.215. For the rank sum test, we found $z = 0.69$.

$$p\text{-value } P(Z > 0.69) = 0.2451$$

9.34   **a.** The output indicates that $t = 3.15$, with a two-tailed p-value of 0.0092.

R.R.:   For one-tailed $\alpha = 0.1$, df $= 11$, reject H_0 if $t > 1.363$

Conclusion: Reject H_0 . The hypothesis that variety R has a higher mean yield is

supported by the data.

b. The "wTest" output shows a Wilcoxon statistic of 70.0 and a two-tailed p-value of 0.017. We should cut the p-value in half for a one-sided research hypothesis. Even as it is, the p-value is much less than $\alpha = 0.10$, so we have support for the research hypothesis.

9.35   The plot of the difference data is given below.

```
-2 | 3
-1 | 7
-0 | 9
 0 |
 1 | 1
 2 |
 3 | 2
 4 | 2 3 6 7
 5 |
 6 | 1
 7 | 9
 8 |
 9 | 7
```

The distribution of differences appears heavy tailed although it has the central mode characteristic of normality. In the case of heavy tails, the signed-rank test may be more powerful. In this situation, since the conclusions of both tests are the same, one test need not be chosen.

9.36   Both *p*-values in the output are for two-tailed tests. Given our one-sided research hypothesis, and the fact that the data *do* appear to go in the direction predicted by H_a, we should divide the *p*-values by 2. Thus we would report $p = 0.0046$ for *t*, and 0.0085 for the signed rank test.

9.37   Since the many plots on a grower's farm may have a wide range in yields (due to fertility, irrigation, etc.), a paired sample experiment eliminates the effect of this variability among plots and concentrates on the variability of the two varieties of trees with which we are concerned. Selecting plots: Choose a wide range (rocky, smooth, fertile, poor mineral content, well irrigated, dry, shady, etc.) so that as many situations as possible are represented in the sample. In this way, one can see how the two varieties compare under many different conditions.

9.44   **a.** The output shows `t = 0.09`, which is much smaller than tabled *t* values for $43 + 26 - 2 = 67$ degrees of freedom (or any degrees of freedom, it doesn't matter in this case). Also, the two-tailed *p*-value is shown as `0.9256`; this *p*-value is much larger than any conventional α. Therefore we can't come close to rejecting the null hypothesis of equal means. The difference is not statistically significant.

b. There certainly is no evidence that the *means* differ. That doesn't prove that the means are the same, only that we have little to no evidence of a real difference. Equally critically, means are not the only interesting question. The labs may differ in variability, skewness, or other statistical properties.

9.45   **a.** The *p*-value is shown as "`Pr > F = 0.0000`." Thus the *p*-value must be some number that is 0 to four decimal places. It must be less than 0.00005.

b. In this situation, we do not want variability around the prescribed value. The higher the variability, the poorer the quality. The result conclusively indicates that source 1 has lower variability, hence better quality.

9.51   **a.** Here are Stata stem-and-leaf displays.

```
. stem AmtSpent if Product==1

            Stem and leaf plot for AmtSpent

         1 | 4556677899
         2 | 00222345566677888889999
         3 | 002333566678889
         4 | 0123368899
         5 | 23566899
         6 | 11267
         7 | 77

. stem AmtSpent if Product==2

Stem and leaf plot for AmtSpent

         1 | 788
         2 | 22233455688999
         3 | 033345667899
         4 | 111122233455566667889
         5 | 0022346778
         6 | 146
         7 | 33
         8 | 8
         9 |
         1 || 08
```

Both plots are somewhat right skewed, not exactly normal. The skewness is not severe.

b. Intervals using pooled-variance t and t' were obtained using Execustat.

```
          Two Sample Analysis for AmtSpent by Product

                                 1                2

Sample size                      73               67
Mean                             36.7736          42.7267     diff.  = -5.95315
Variance                         247.67           271.673     ratio  = 0.911647
Std. deviation                   15.7375          16.4825

95% confidence intervals
     mu1 - mu2: (-11.3385,-0.567797) assuming equal variances
     mu1 - mu2: (-11.3501,-0.556203) not assuming equal variances
     variance ratio: (0.564762,1.46416)
```

In both cases, the interval does not include 0.00, so that we can reject (at $\alpha = 0.05$) the null hypothesis of equal means.

c. The intervals are almost exactly equal. The reason is that the sample standard deviations were almost equal, as were the sample sizes. In such a case, the question of pooled-variance t versus t' is immaterial.

9.52   **a.** The test results using Stata follow.

```
. ttest AmtSpent,  by(Product)

Variable |     Obs        Mean     Std. Dev.
---------+------------
       1 |      73     36.77356     15.73753
       2 |      67     42.72672      16.4825
---------+------------
combined |     140     39.62257     16.31541

        Ho:  mean(x) = mean(y)   (assuming equal variances)
                    t = -2.19 with 138 df
              Pr > |t| = 0.0305

. ttest AmtSpent,  by(Product) unequal

Variable |     Obs        Mean     Std. Dev.
---------+------------
       1 |      73     36.77356     15.73753
       2 |      67     42.72672      16.4825
---------+------------
combined |     140     39.62257

        Ho:  mean(x) = mean(y)   (assuming unequal variances)
                    t = -2.18 with 138 df
              Pr > |t| = 0.0308
```

In each case, the p-value is 0.031, which is less than $\alpha = 0.05$, so we may reject H_0 .

b. In Stata, the rank sum test is called a test of equality of medians.

```
. ranksum AmtSpent,  by(Product)

Test: Equality of medians (Two-Sample Wilcoxon Rank-Sum)

Sum of Ranks: 5264 (Product == 2)
Expected Sum: 4723.5

z-statistic   2.25
Prob > |z|    0.0242
```

The p-value is shown as 0.0242.

c. The rank sum test is very slightly more conclusive than either t test, but all three give essentially the same answer. The assumptions for all three tests seem to be met reasonably well, so all three are believable.

Chapter 10

Methods for Proportions

10.1   Two-Sample Procedures for Proportions

10.4   **a.** The 95% confidence interval for the difference in proportions is

$$\hat{\pi}_1 - \hat{\pi}_2 - z_{\alpha/2}\sigma_{\hat{\pi}_1\hat{\pi}_2} \le \pi_1 - \pi_2 \le \hat{\pi}_1 - \hat{\pi}_2 + z_{\alpha/2}\sigma_{\hat{\pi}_1\hat{\pi}_2}$$

where

$$\alpha = 0.05 \;\text{ so that }\; z_{\alpha/2} = 1.96$$

$$\hat{\pi}_1 = \frac{674}{983} = 0.6857 \qquad \hat{\pi}_2 = \frac{1{,}697}{2{,}961} = 0.5731$$

$$\hat{\pi}_1 - \hat{\pi}_2 = 0.1126$$

and

$$\sigma_{\hat{\pi}_1\hat{\pi}_2} = \sqrt{\frac{\hat{\pi}_1(1-\hat{\pi}_1)}{n_1} + \frac{\hat{\pi}_2(1-\hat{\pi}_2)}{n_2}} = \sqrt{\frac{0.6857(0.3143)}{983} + \frac{0.5731(0.4269)}{2{,}961}} = 0.0174$$

Therefore,

$$0.1126 - 1.96(0.0174) \le \pi_1 - \pi_2 \le 0.1126 + 1.96(0.0174)$$
$$0.1126 - 0.0341 \le \pi_1 - \pi_2 \le 0.1126 + 0.0341$$
$$0.0785 \le \pi_1 - \pi_2 \le 0.1467$$

Note: $n_1\hat{\pi}_1$, $n_1(1-\hat{\pi}_1)$, $n_2\hat{\pi}_2$, $n_2(1-\hat{\pi}_2)$ are all greater than 5, so use of the normal approximation should be reasonably accurate.

b. We first note that we have a directional (one-sided) research hypothesis, that a higher percentage in population 1 are regular watchers. The confidence interval is two-sided, so there is a slight technical difficulty in using it for this test. Instead, we will go through the familiar formal test.

H_0: $\quad\quad\quad \pi_1 - \pi_2 = 0$

H_a: $\quad\quad\quad \pi_1 - \pi_2 > 0$

T.S.: $\quad\quad z = \dfrac{\hat{\pi}_1 - \hat{\pi}_2 - 0}{\sigma_{\hat{\pi}_1 \hat{\pi}_2}} = \dfrac{0.1126}{0.0174} = 6.47$

R.R.: $\quad\quad$ For $\alpha = 0.05$, reject H_0 if $z > 1.645$

Conclusion: Reject H_0. The data support a hypothesis that a higher percentage of rural
$\quad\quad\quad\quad\quad\quad$ residents regularly watch the news progeam.

Practically speaking, the confidence interval for the difference in proportions was so far from including 0.00 that we clearly could support our research hypothesis, disregarding the one-sided or two-sided issue.

10.5 Recall that the p-value is the tail area beyond the actually observed test statistic. We need to know the choice of test statistic (Z, in this case), its actual value (6.47, in this case), and whether we want a one-tailed or two-tailed value (one, in this case).

$$\text{p-value} = P(Z > 6.47) = \text{an } \textit{extremely} \text{ small number} \approx 0.$$

The results of the study strongly support the research hypothesis.

10.6 This exercise illustrates the device of testing a difference of proportions by coding successes as 1's, failures as 0's. A *t* test of the means of these 1's and 0's is almost identical to the *z* test of a difference of proportions outlined in the text. The output, under the "Hypothesis Test - Difference of Means" heading, shows a *t* statistic equal to 2.30, and a *p*-value equal to 0.0218. Because the *p*-value is less that $\alpha = 0.05$, we can support the research hypothesis. In nontechnical language, we have fairly strong evidence that the difference in proportions in the sample is more than just random variation, the difference reflects a real difference in the underlying populations.
$\quad\quad$ For comparison, we can calculate the *z* statistic. The sample proportions are simply the means of the coded 1's and 0's; they are shown as 0.855 for sample 1 and 0.765 for sample 2. We have $n_1 = n_2 = 200$, $\hat{\pi}_1 = 0.855$, and $\hat{\pi}_2 = 0.765$. H_0: $\pi_1 - \pi_2 = 0$ and H_a: $\pi_1 - \pi_2 \neq 0$. The test statistic is

$$z = \frac{(\hat{\pi}_1 - \hat{\pi}_2) - 0}{\sqrt{\frac{\hat{\pi}(1-\hat{\pi})}{n_1} + \frac{\hat{\pi}(1-\hat{\pi})}{n_2}}} = \frac{(0.855 - 0.765) - 0}{\sqrt{\frac{0.855(0.145)}{200} + \frac{0.765(0.235)}{200}}} = 2.31$$

This value is, for all practical purposes, equal to the t statistic shown in the output.

10.7 The 95% confidence interval shown for the difference of means "mu1 - mu2" can be interpreted as applying to the difference of proportions. The mean of a sample of 1's and 0's is nothing more than the proportion of 1's in the sample. The output lists two different intervals, depending on whether or not we assume equal variances. Technically, we're not making that assumption, but it doesn't matter at all in this case. The intervals start differing in the fifth decimal place!

$$0.013 \le \pi_1 - \pi_2 \le 0.167$$

The 95% confidence interval doesn't include 0.000, so once again we reject H_0 at $\alpha = 0.05$.

10.10 **a.** We assume that samples of 30 motors are tested for each supplier. Thus $n_1 = 30$, $\hat{\pi}_1 = 22/30 = 0.733$, $n_2 = 30$, and $\hat{\pi}_2 = 16/30 = 0.533$. The test statistic is

$$z = \frac{(\hat{\pi}_1 - \hat{\pi}_2) - 0}{\sqrt{\frac{\hat{\pi}(1-\hat{\pi})}{n_1} + \frac{\hat{\pi}(1-\hat{\pi})}{n_2}}} = \frac{(0.733 - 0.533) - 0}{\sqrt{\frac{0.733(0.267)}{30} + \frac{0.533(0.467)}{30}}} = 1.64$$

We note that $|z| \le z_{0.05/2} = 1.96$. H_0 must be retained so the difference is not statistically significant.

b. To say that the two suppliers provide equally reliable motors is to say that H_0 is true. We can't prove that H_0 is true, most especially with small sample sizes. We can only retain H_0 as a possibility. We have not proved that the suppliers differ; that's different than saying we've proved they are exactly the same.

10.11 The confidence interval is

$$(\hat{\pi}_1 - \hat{\pi}_2) - z_{\alpha/2}\sqrt{\frac{\hat{\pi}_1(1-\hat{\pi}_1)}{n_1} + \frac{\hat{\pi}_2(1-\hat{\pi}_2)}{n_2}} \le \pi_1 - \pi_2 \le (\hat{\pi}_1 - \hat{\pi}_2) + z_{\alpha/2}\sqrt{\frac{\hat{\pi}_1(1-\hat{\pi}_1)}{n_1} + \frac{\hat{\pi}_2(1-\hat{\pi}_2)}{n_2}}$$

$$(0.733 - 0.533) - 1.96\sqrt{\frac{0.733(0.267)}{30} + \frac{0.533(0.467)}{30}} \le \pi_1 - \pi_2$$

$$\le (0.733 - 0.533) + 1.96\sqrt{\frac{0.733(0.267)}{30} + \frac{0.533(0.467)}{30}} \quad -0.039 \le \pi_1 - \pi_2 \le 0.439$$

The interval is extremely wide. We are 95% confident that supplier 1 produces between 3.9% fewer reliable motors and 43.9% more reliable motors than does supplier 2! Thus we know very little.

10.2   Tests for Several Proportions

10.14   **a.**   To calculate the expected frequencies, use the fact that

$$E_i = n\pi_i \text{ where } n = 25$$

The results are shown in a table in the next part of the answer.

b.   The following table summarizes the information needed for the desired goodness of fit test.

Number of arrivals per minute	Theoretical Proportions π_i	Expected Frequencies $n\pi_i$	Observed Frequencies n_i	$n_i - E_i$
0	0.30	7.50	10	+2.50
1	0.45	11.25	6	−5.25
2 or more	0.25	6.25	9	+2.75
	1.00	25	25	0

The goodness of fit test can be written out in five steps.

H_0 : $\pi_i = \pi_{i,\,0}$ (as specified in the above table)

H_a : H_0 is not true

T.S.: $\chi^2 = \sum_{i=1}^{3} \dfrac{(n_i - E_i)^2}{E_i} = \dfrac{(2.5)^2}{7.5} + \dfrac{(-5.25)^2}{11.25} + \dfrac{(275)^2}{6.25} = 4.4933$

R.R.:   For $\alpha = 0.05$, the χ^2 table value, 2 df is 5.991. Reject H_0 , that the shop owner's theoretical proportions are correct if $\chi^2 > 5.991$

Conclusion: The null hypothesis cannot be rejected at the 0.05 level.

Technically, there is "a good fit" to the data. However, since n is small and since all of the expected frequencies are not much larger than 5, the probability of a Type II error, false acceptance of the null hypothesis, may be very high.

10.15   **a.** The following table summarizes the information needed for the desired goodness of fit test

Number of Purchases	Theoretical Proportions π_i	Expected Frequencies $n\pi_i$	Observed Frequencies n_i	$n_i - E_i$
0	0.30	75.0	100	+25.0
1	0.45	112.5	60	−52.5
2 or more	0.25	62.5	90	+27.5
	1.00	250	250	0

Use the goodness of fit test.

H_0: $\pi_i = \pi_{i,0}$ (as specified in the above table)

H_a: H_0 is not true

T.S.: $\chi^2 = \sum_{i=1}^{3} \dfrac{(n_i - E_i)^2}{E_i} = \dfrac{(25)^2}{75} + \dfrac{(-52.5)^2}{112.5} + \dfrac{(27.5)^2}{62.5} = 44.9333$

R.R.:   For $\alpha = 0.05$, df $= 2$, reject H_0 if $\chi^2 > 5.99$.

Conclusion: Reject H_0. The data do not support the shop owner's theoretical distribution.

b. This conclusion differs from the previous one, despite the fact that the sample proportions were exactly the same (0.50, 0.20, and 0.30) in both cases. Here the sample size is larger, and so the expected cell frequencies are all proportionally larger. For these reasons, the probability of a type II (false negative) error is small compared to the previous situation, where it was very likely that the null hypothesis would not be rejected when it was false. Note that when the sample size and observed frequencies were multiplied by ten, the calculated χ^2 was also multiplied by ten.

10.3   Chi-Square Tests for Count Data

10.18   **a.** The expected counts are not shown in the output. To find them, use the fact that

$$\hat{E}_{ij} = \frac{n_i n_j}{n}$$

For example, to find the expected number of middle managers who are age 30–39 and not promoted:

$$\hat{E}_{22} = \frac{170(70)}{250} = 47.6$$

The table of expected numbers is given below:

	Promoted	Not Promoted	total
Under 30	16.0	34.0	50
30–39	22.4	47.6	70
40–49	25.6	54.4	80
Over 50	16.0	34.0	50
Total	80.0	170.0	250.0

b. The degrees of freedom depend on the number of rows and columns in the table, not on the sample size.

$$df = (r-1)(c-1) = 2(2-1)(4-1) = 1 \times 3 = 3$$

c. The chi-square statistic is shown as 13.025. We must compare it to the appropriate χ^2 table value from Appendix Table 5. The entry for $df = 3$ and right tail area 0.05 is 7.815.

R.R: For $\alpha = 0.05$, $df = 3$, reject H_0 if $\chi^2 > 7.815$

Conclusion: Reject H_0. The data support the hypothesis that age and promotion status
are related.

10.19 To find a p-value, we need the actually observed statistic (13.025, in this case), the table to use (Appendix Table 5), and an indication of whether we want a one-tailed or two-tailed area. In χ^2 tests, the rejection region is only the upper tail, so we want a one-tailed p-value. Checking Table 5 with 3 df, we find that 13.025 falls between 12.84 (area 0.005) and 16.27 (area 0.001). The area beyond 13.025 must be somewhere between 0.001 and 0.005.

$$0.001 < p\text{-value} = P\left(\chi^2 > 13.0252\right) < 0.005$$

10.20 **a.** The chi-square value is shown as a very small value, 0.012, in the output. With 2 rows and 2 columns, there is 1 degree of freedom.

R.R.:   For $\alpha = 0.05$, df $= 1$, reject H_0 if $\chi^2 > 3.84$

Conclusion: Do not reject H_0 . The data are consistent with (but don't prove) the null hypothesis that the variables are independent.

b. In Exercise **10.18**, when we were dealing with four age groups, the data strongly supported the research hypothesis that some relationship existed between age and promotion status. In part **a.** of this exercise, combining age groups when there was no need to (all cells sufficiently larger than 5) had the effect of completely masking a relationship which was present. The combined categories "hide" the fact that very few middle managers under 30 or over 50 are promoted compared to the 30–50 age range.

Supplementary Exercises

10.25   **a.** Under the hypothesis of independence, the expected frequencies are given in the following Excel table:

		Opinion				
Commercial	1	2	3	4	5	Total
A	42	107	78	34	39	300
B	42	107	78	34	39	300
C	42	107	78	34	39	300
Total	126	321	234	102	117	900

Notice that because there are 300 consumers in each commercial group, the expected frequencies are the same within each opinion category.

b. There are $(5 - 1)(3 - 1) = 8$ degrees of freedom for testing the hypothesis of independence.

c. The cell chi-squares are summarized below:

	Opinion				
Commercial	1	2	3	4	5
A	2.3810	3.7383	2.1667	4.2353	0.6410
B	2.8810	10.8037	0.0513	5.7647	21.5641
C	0.0238	1.8318	1.5513	0.1176	14.7692

The overall $\chi^2 = 72.521$ with 8 degrees of freedom. The p-value is 0.0000 (effectively 0). Therefore we can reject the hypothesis of independence and conclude that there is an association between which commercial was viewed and opinion.

10.26   We want a one-tail p-value because the test rejects H_0 only for values in the right tail.

$$p\text{-value} = P\left(\chi^2 > 72.521\right)$$

For df $= 8$, $\chi^2 = 26.12$ is the largest value in the table, and this value corresponds to a tail area of 0.001. Since the calculated χ^2 value is much larger than this, we know that the p-value is much less than 0.001. In English this means that if in fact there is no relationship, the probability of obtaining the sample results that we did due to random fluctuation is extremely small.

Chapter 11

Analysis of Variance and Designed Experiments

11.1 Testing the Equality of Several Population Means
11.2 Comparing Several Distributions by a Rank Test

11.6   **a.** The means and standard deviations are shown in the output.

Policy	Mean	Standard deviation	Sample size
1	7.9437	3.4875	16
2	2.7600	2.1879	10
3	7.5500	5.2944	10

The grand mean is the shown as the TOTAL mean, 6.0845.

b. The degrees of freedom and mean square (MS) values are shown in the output in the section labeled ONE-WAY AOV FOR WAIT BY POLICY. The df are 2 (between) and 33 (within), the MS are 93.6535 (between) and 14.4790 (within).

c. The F statistic is the ratio of the mean squares found in the previous exercise. It is shown in the output under ONE-WAY AOV as F = 6.47.

d. We should, ideally, compare the computed F value to tabulated F value with 2 and 33 df; we don't have such values. The tabled F value for 2 and 33 df and $\alpha = 0.01$ must be between the values for 2 and 30 df (5.39) and for 2 and 40 df (5.18). Roughly, the tabled value is about 5.3. Whatever the exact table value may be, the computed F statistic clearly is larger. Therefore, we may reject H_0.

e. The p-value is shown in the output, directly following the F statistic, as 0.0043. Because this is less than $\alpha = 0.01$, once again we may reject the null hypothesis.

11.7   **a.** Most times should be reasonably short, but occasionally there will be long waiting times. These occasional long times should give a right skew to the data. This seems to be the basic pattern in the stem-and-leaf displays. The data are not normally distributed.

b. This skewness would not completely invalidate the test. The Central Limit Theorem applies, because we are analyzing sample means. Although the total sample size, 36, isn't very large, the results should be a fair (not excellent) approximation.

11.8   **a.** A Minitab time series plot of the data is shown below. We don't see any clear trend, either increasing or decreasing. Nor is there any strong cyclic effect; there may be lows around 20, and 30, but nothing we would call evident.

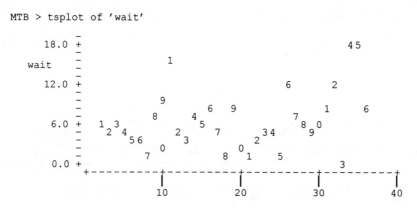

```
MTB > tsplot of 'wait'
```

b. A crucial assumption for analysis of variance, as well as for most other statistical methods, is that the observations are independent. That assumption implies that there should be no pattern in a plot of data against time; the data should be random. If there is a pattern, the assumption of independence would be falsified.

11.9   **a.** The output shows H $= 14.57$ and a p-value of 0.0007. Because the p-value is so small, we have a clearly detectable (significant) difference among groups.

b. By both tests, we found that there was a detectable difference. The evidence was quite conclusive from both tests; the p-value for the Kruskal-Wallis test is less than 0.001 and the p-value for the F test is about 0.004. The Kruskal-Wallis test is slightly more conclusive, but the results are very similar.

11.16   **a.** The computed F statistic is shown as 0.586. If H_0 is true, the expected value of F is about 1.00, so the computed F is not even as large as the expected value. Certainly it is not large enough to indicate that there are statistically detectable differences among means. All tabled F values are substantially larger than 1, so the result is not significant, no matter what α is chosen. Another way to reach the same conclusion is to note that the p-value is 0.5575, much larger than conventional α values.

b. There is no evidence at all to indicate that the average sale amount differs among the order form types. This doesn't absolutely prove that there is no difference; it does say that the study provides no evidence of a real difference. The *p*-value is 0.586, far larger than any normal α value.

11.17 **a.** The Empirical Rule clearly doesn't work for these data. We might suspect outliers in both directions, but clearly we can't ever have a very large negative order. The only plausible explanation we can think of is right skewness. This makes sense in context; most orders will be small, but a few will be quite large. Therefore, it seems likely that order amounts data will be right skewed.

b. The normal-distribution assumption seems unreasonable, given the likely skewness. We should note that the Central Limit Theorem effect means that the claimed probabilities should be decent approximations. However, the ANOVA procedure may not be the most efficient.

11.18 The *p*-value for the Kruskal-Wallis test is 0.210. As we suspected, the Kruskal-Wallis test is more efficient for the skewed data here. However, a *p*-value of 0.210 is still not conclusive evidence that there is a real, more-than-random difference among means.

11.3 Specific Comparisons Among Means

11.20 The output indicates that there are two groups in which the means do *not* differ significantly. The "REJECTION LEVEL 0.050" entry indicates that the computation used $\alpha = 0.05$. It appears that policies 1 and 3 form one group; policy 2, the other. If we look at the means, this makes sense. The means for policies 1 and 3 are quite similar, both in the 7's. The mean for policy 2 is much smaller. So we can translate the output as indicating that the mean for policy 2 is significantly ($\alpha = 0.05$) different than the means for policies 1 and 3, but the latter two means are not significantly different.

11.21 The problem with haphazard selection of a policy for the day is that the store manager may unconsciously bias the selection. A favored policy might be chosen for days that were anticipated to be easy or a disliked policy might be chosen for bad days. To avoid such biases, the policy should be chosen randomly. Ideally, there should be equal numbers of measurements for each policy. To achieve a random selection for 36 days, we could put 12 slips of paper numbered 1, 12 numbered 2, and 12 numbered 3 into a box, and draw one for each day. The same effect could be achieved by using computer-generated random numbers.

11.4 Two-Factor Experiments

11.27 **a.** To calculate the grand mean, find the average of all 12 cell means.

$$\text{Grand Mean} = \frac{1}{12}(2.1 + 1.7 + 1.3 + 1.7 + 1.9 + 1.8 + 1.7 + 1.8 + 0.8 + 0.7 + 0.6 + 0.7) = 1.4$$

Alternatively, we could find the mean for each row (each plan) and average those. Or find the mean for each column (each area) and average those. The result is the same.

b. To calculate row effects, calculate row means and subtract the grand mean from each row mean. The row effects are calculated below.

Plan	Row Mean	Row Effect
A	1.7	1.7 – 1.4 = 0.3
B	1.8	1.8 – 1.4 = 0.4
C	0.7	0.7 – 1.4 = –0.7

To calculate column effects, subtract the grand mean from each column mean. The column effects are calculated below.

Area	Column Mean	Column Effect
1	1.6	1.6 – 1.4 = 0.2
2	1.4	1.4 – 1.4 = 0
3	1.2	1.2 – 1.4 = –0.2
4	1.4	1.1 – 1.4 = 0

In passing, note that the row effects add to 0.0, as do the column means. The sum of deviations from a mean is always 0.

11.28 Exercise **11.27** gives the mean square (MS) values, as well as the relevant degrees of freedom. To calculate F statistics, all we have to do is divide each MS by MS(Error). For example, for Plan, $F = \text{MS(Plan)}/\text{MS(Error)} = 29.60/120 = 24.67$. We could compare each computed F value to the tabled value for those df and $\alpha = 0.05$. For example, compare the F statistic for Plan to the tabled value for 2 and 228 degrees of freedom. We don't have 228 df, so we'll use the next lower df shown, namely 120. The 0.05 value is 3.07. If we had used 2 and 240 df, the value is 3.03, so it hardly matters which we use. The test statistic for Plan is much larger than the tabled value, so we would declare the `Plan` effect significant. Alternatively, note that F for `Plan` is larger than even the 0.001 value in the table. Therefore we would reject the null hypothesis even for $\alpha = 0.001$. By the Universal Rejection Region, this means that the p-value is less than 0.001. We proceed in the same way for `Area` and `Interaction`. One convenient way to summarize the results is to complete the ANOVA table.

Source	SS	df	MS	F	p-value
Plan	59.2	2	29.6	24.67	$p < 0.001$
Area	4.8	3	1.6	1.33	$p > 0.25$
Interaction	2.4	6	0.4	0.33	$p > 0.25$
Error	273.6	228	1.2		
Total	340.0	239			

The conclusions of the F tests are summarized below. The tabled values are approximations between the entries for 120 df and 240 df; in every case the test statistic is far from the table value, so the exact table value isn't important.

Null hypothesis	F	Conclusion	
c. No interaction	0.33	$F < F_{6,\ 228} = 2.14$	Retain H_0
a. No plan (row) effect	24.67	$F > F_{2,\ 228} = 3.04$	Reject H_0
b. No area (col.) effect	1.60	$F < F_{3,\ 228} = 2.65$	Retain H_0

The data indicate that only *Plan* has a serious effect on differences in one-year increase in income.

11.29 **a.** Remember that a profile plot has means on the vertical axis, the levels of one factor on the horizontal axis, and the levels of the other factor used to connect points. A profile plot for the data is shown below.

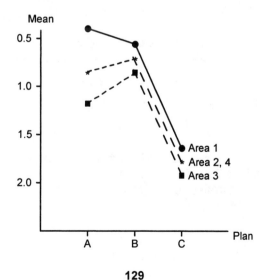

b. The plot suggests the possibility that some interaction is present. The profiles are not perfectly parallel. The differences between plans (in average one-year increase in income) are not the same for all areas.

c. The F test of Exercise **11.28** indicates that it is not possible to reject the null hypothesis that no interaction is present. This appears to conflict with our profile plot. Although the plot appears to indicate some interaction, this departure from parallelism is not enough for statistical significance; that is, it could reasonably be random variation.

11.30 The Tukey 95% confidence interval for comparison of row means is calculated by taking the difference of row means ± the appropriate value from Appendix Table 8 times the square root of MS(Error) divided by the sample size per mean. That is,

$$\left(\bar{y}_{i..} - \bar{y}_{i'..}\right) - q_{0.05}(I,\ df_2)\sqrt{\frac{MS(Within)}{nJ}} \le \mu_{i..} - \mu_{i'..} \le \left(\bar{y}_{i..} - \bar{y}_{i'..}\right) + q_{0.05}(I,\ df_2)\sqrt{\frac{MS(Within)}{nJ}}$$

where

$$n = 20$$
$$I = 3 \qquad\qquad\qquad \bar{y}_{1..} = 1.7$$
$$J = 4$$
$$MS(Within) = 1.2 \qquad\qquad \bar{y}_{2..} = 1.8$$
$$df_2 = 228 \approx \infty$$
$$\bar{y}_{3..} = 0.7$$
$$q_{0.05}(3,\ \infty) = 3.31$$

Therefore,

$$-0.1 - 3.31\sqrt{\frac{1.2}{20(4)}} \le \mu_1 - \mu_2 \le -0.1 + 3.31\sqrt{\frac{1.2}{20(4)}}$$
$$-0.51 \le \mu_1 - \mu_2 \le 0.31$$
$$1 - 3.31(0.1225) \le \mu_1 - \mu_3 \le 1 + 3.31(0.1225)$$
$$0.59 \le \mu_1 - \mu_3 \le 1.41$$
$$1.1 - 3.31(0.1225) \le \mu_2 - \mu_3 \le 1.1 + 3.31(0.1225)$$
$$0.69 \le \mu_2 - \mu_3 \le 1.51$$

According to the Tukey confidence intervals, the mean one-year increase in income for Plan A is significantly different from that of Plan C. Also, the mean for Plan B is significantly different from that of Plan C. However, the means for Plans A and B are not significantly different.

11.31   **a.** In any two-factor study, the first thing to check is the possible presence of interaction. The table of means shown in the output shows that mixture 1 has the second *highest* score at altitude 1, and a substantially higher mean than mixture 2 at altitude 2. Mixture 1 has by far the lowest score at altitude 3, but that's where the fewest customers live. There is a major interaction in the data; we must look beyond the overall averages.

b. For altitude 1, the profile decreases from mixture 1 to mixture 2 to mixture 3, then increases for mixture 4. For altitude 2, the profile decreases from mixture 1 to mixture 2, then increases to mixture 3 and again to mixture 4. The profile for altitude 3 starts very low at mixture 1, and jumps way up for mixtures 2, 3, and 4. The profiles aren't close to parallel. They cross each other. This fact indicates that there is a substantial mixture-altitude interaction.

11.32   **a.** The simplest way to see that there is a detectable interaction is to calculate an F statistic for the altitude-mixture interaction.
$F = \text{MS(Interaction)}/\text{MS(Error)} = 1944/334 = 5.82$. Compare this number to the F table (Appendix Table 6) with 6 and 60 df. The largest value in the table, corresponding to $\alpha = 0.001$, is 4.37. Therefore, we reject the hypothesis of no interaction, even at $\alpha = 0.001$, so the p-value is something less than 0.001. Because the p-value is smaller than any reasonable α, by the Universal Rejection Region, we can claim a statistically detectable interaction.

b. Interaction means that the difference between mixtures in average values depends on the altitude. Thus, the relative desirability of mixtures, and particularly which mixture is optimal, depends on the altitude at which it is used. We saw in the previous exercise that mixture 1 was optimal at altitude 3, but not at the lower altitudes.

c. When there is serious interaction, the main effects aren't very meaningful. Perhaps if the three different altitudes contained roughly equal numbers of customers, and if the same mixture had to be used for all altitudes, the mixture averages would be sensible measures. But that's not true here; altitude 3 has the fewest customers. Thus the test for the main effect of mixtures is quite misleading.

11.5   Randomized Block Experiments

11.38   **a.** The factor of most interest is presumably the `mixture` factor; the `investigator` factor is a "nuisance" factor. Thus the treatments are `mixtures` and the blocks are `investigators`.

b. The randomized block design controls for any systematic variation among investigators. If some investigators tended to measure the thrust higher on average than did other investigators, there would be variation among investigators. In a completely randomized design, any variation among investigators would inflate the error SS.

11.39 **a.** The F test for differences among investigators is not statistically significant. The p-value is 0.2273, and the F statistic is small at 1.64. However, the F statistic for differences among mixtures is enormous; $F = 1,264.7$. This value is far beyond any value in the F table, and certainly is significant. The output shows a p-value with fourteen 0's after the decimal point.

b. It seems reasonable that the higher the response (propellant thrust), the better. Therefore, mixture 2, with a mean of 2,653.2, appears much better than any of the other mixtures, which all have means of roughly 2,400. To compare pairs of means, we note that each mixture mean is based on $n = 5$ measurements, that the mean square for error is 68.858, and that there are 12 error df. The Tukey table value (for $\alpha = 0.05$, 4 means, 12 df) is 4.20. Thus the $\pm$ for confidence intervals is

$$4.20\sqrt{\frac{68.858}{5}} = 15.59$$

The difference between the mixture 2 mean and any other mean is more than 100, so any confidence interval comparing mixture 2 with any other mixture surely won't include 0. In fact, all pairwise differences of means are statistically significant.

11.40 No; the p-value for INVEST is large at 0.2273. The differences among inverstigator averages might have arisen by random variation. If the research hypothesis of nonzero investigator effects had been supported, there would be evidence of systematic differences among investigators in the average measurement of thrust.

Supplementary Exercises

11.51 **a.** The F statistic for the null hypothesis of equal means is shown in the "Analysis of Variance" part of the output as F = 23.10.

b. H_0: $\qquad \mu_1 = \mu_2 = \mu_3$

H_a: $\qquad$ Not all means equal

T.S.: $\qquad F = 23.10$

R.R.: $\qquad$ For $\alpha = 0.01$, reject H_0 for $F > F_{0.01,\, 2,\, 29} = 5.42$

Conclusion: Reject H_0. The data support the hypothesis of unequal means.

c. The p-value for this test is shown as `0.0000` and therefore is some number less than 0.00005. The research hypothesis of unequal means is strongly supported.

11.52 **a.** The Kruskal-Wallis statistic is shown in the "`kwallis`" part of the output, as `chi-square = 18.843`.

b. We certainly could look up the value of the Kruskal-Wallis statistic in the chi-squared tables (Appendix Table 5). However, the easier way to respond to the question is to use the p-value shown in the output as 0.0001. Because the p-value is (much) smaller than $\alpha = 0.01$, we may clearly reject the hypothesis of equal locations.

11.53 **a.** There are several ways to plot the data. Here are stem-and-leaf displays, for example.
Commercial:

Clothes	Food	Toys
1 | 8	1 |	1 |
2 | 1 1 3	2 |	2 |
2 |	2 | 5	2 |
3 | 0 0	3 | 0 2	3 |
3 | 6 7	3 | 8	3 |
4 | 2	4 | 1 1	4 |
4 |	4 | 5 6	4 | 7 8 9 9
5 |	5 | 0 1 3	5 | 1 2
5 |	5 | 7	5 | 5 6 8 9
6 |	6 |	6 | 0

There isn't any obvious skewness, nor any outliers.

b. There appear to be very slight violations, if any, of the ANOVA assumptions. The three plots are not seriously skewed nor outlier-prone. The standard deviations and sample sizes are somewhat different. The combination of unequal variability and unequal sample sizes does make the F probabilities somewhat inaccurate. However, the F statistic is so far beyond table values that there is no reason to worry about such minor difficulties. There is no indication of bias (though some may be present) nor of dependent observations.

c. Since both the *F* test and the Kruskal-Wallis test show *strong* support of the research hypothesis (both *p*-values much smaller than even 0.001) choice of the more appropriate test is not at all important here.

11.54 The problem here is that the sample sizes are not the same. Therefore, the calculations given in the text aren't strictly appropriate. The text indicates that most computer packages will handle unbalanced sample sizes for the Tukey procedure. Here, for example, is how Minitab handled the problem.

```
MTB > Oneway 'Attspan' 'Commtype';
SUBC>    Tukey 1.

One-Way Analysis of Variance

Analysis of Variance on Attspan
Source      DF        SS        MS        F        p
Commtype     2    2953.6    1476.8    23.10    0.000
Error       29    1853.8      63.9
Total       31    4807.5
                                        Individual 95% CIs For Mean
                                        Based on Pooled StDev
    Level      N      Mean     StDev    ----+-------+-------
        1      9    28.667     8.426    (-----*----)
        2     12    42.417     9.839                (---*----)
        3     11    53.091     4.700                       (---*---)
                                        -----+-------+------
    Pooled StDev =     7.995              30      40        50

Tukey's pairwise comparisons

        Family error rate = 0.0100
    Individual error rate = 0.00368

    Critical value = 4.47

    Intervals for (column level mean) - (row level mean)

                    1          2

        2      -24.89
               -2.61

        3      -35.78     -21.22
               -13.07      -0.13
```

All the intervals shown do not include 0.00 in the confidence interval. Therefore, all the differences are statistically detectable at $\alpha = 0.01$.

11.55 **a.** The null hypothesis of equality of house prices is totally irrelevant here. Only if all the houses were built exactly alike by one contractor with similar locations would we expect house prices to be equal. Since no description of the houses was provided we assume this is not the case. The *F* test concurs: $F = 122.58$ has a *p*-value $= 0.0000$ (less than 0.00005).

b. The *F* statistic for the null hypothesis of equality of appraiser effects is shown as $F = 10.66$ ($df_1 = 3$, $df_2 = 33$).

c. This null hypothesis is rejected at all typical α levels. The p-value for this test is shown as 0.0000, which means it is less than 0.00005. There is clear evidence that the appraisers differ in their average evaluation.

    Note: The problem extends the idea of paired samples seen in Chapter 9. The great variability in house prices can be controlled by matching each house to its four appraised values.

11.56   The output shows two "homogeneous groups." Presumably, any mean in one group differs significantly from a mean in the other group. Looking at the means, we note that the mean for appraiser 2 is substantially lower than any of the other means. Mean 2 is in a group by itself. If we read the output correctly, appraiser 2's mean differs significantly from the other means, but the other means do not differ significantly among themselves.

    We can check our presumptions by calculation. The statement of the problem specifies $\alpha = 0.05$. There are 4 means, each based on a sample size of 12, and the MS(Error) is shown in the output is 11.930, with 33 df The required Table 8 value is between 3.85 (30 df) and 3.79 (40 df); conservatively, we can take the value as 3.85. The Tukey ± number is therefore

$$3.85\sqrt{\frac{11930}{12}} = 3.84$$

If a difference between means is larger than 3.84 (in magnitude), the confidence interval for the difference will not include 0.00, and will therefore indicate a statistically significant (detectable) difference. Note that the output shows a critical Q value of 3.826 and a critical value for comparison (the ± number) as 3.8151, using the correct df of 33. Only the difference between the mean for appraiser 2 and any other mean is larger than 3.84 (or 3.8151), and therefore statistically detectable. This is exactly what we presumed from looking at the output.

11.57   **a.** The F statistic for the null hypothesis of equal appraiser means is $F = 0.34$ ($df_1 = 3$, $df_2 = 44$). This is much less than any table value. In fact, the expected value of F assuming that the null hypothesis is true is just about 1.0; here, the F statistic is even smaller than that. To confirm, the p-value is shown as 0.7968, much larger than 0.05. The null hypothesis cannot be rejected at the $\alpha = 0.05$ level.

    b. It is important to control for the effects of house differences for this data. The variation for houses completely masks the variation for appraisers which we are attempting to study, as can be seen in the one-factor analysis of variance. The two-factor analysis allowed us to view variation in appraisers by controlling for the effect of house differences.

11.67   The output shows an F statistic equal to 17.08, and a p-value of 0.000. The actual p-value presumably is some unspecified number less than 0.0005. Whatever the exact p-value may be, it is certainly less than $\alpha = 0.05$. By the Universal Rejection Region, there is a statistically significant difference somewhere among the means.

11.68   The Kruskal-Wallis statistic is shown as `H = 28.32`, with a *p*-value of 0.000 (so less than 0.0005). The result is the same as for the *F* test.

11.69   The desire is to have as large a number of copies as possible, so the higher the mean, the better. Design 3 has the highest mean. All the intervals involving mean 3 do not include 0; therefore, mean 3 is significatly (detectably) different from each of the other means. Here is an unambiguous result; design 3 is clearly the best, beyond random variation.

11.73   **a.** We used Systat to obtain boxplots.

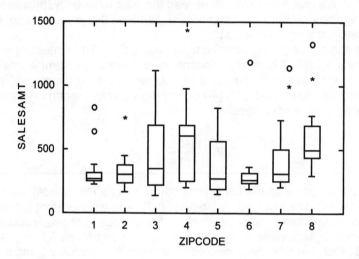

These plots definitely are not normal. In several of the boxplots, there are outliers shown on the high side. Some of the outliers, indicated by the O symbol, are major ones. In addition, virtually all the boxes are right-skewed, with a long tail toward the higher values.

b. Systat's ANOVA table is the following.

```
LEVELS ENCOUNTERED DURING PROCESSING ARE:
ZIPCODE
      1.000        2.000        3.000        4.000        5.000
6.000
      7.000        8.000
```

DEP VAR:SALESAMT   N: 115   MULTIPLE R: 0.328   SQUARED MULTIPLE R: 0.108

ANALYSIS OF VARIANCE

SOURCE	SUM-OF-SQUARES	DF	MEAN-SQUARE	F-RATIO	P
ZIPCODE	870264.730	7	124323.533	1.849	0.085
ERROR	7195856.053	107	67250.991		

136

The p-value is shown as 0.085, so H_0 can be rejected at $\alpha = 0.10$, but not at $\alpha = 0.05$. At best we have modest evidence of a real ZIP code effect.

11.74   **a.** The Systat Kruskal-Wallis test is shown here.

```
KRUSKAL-WALLIS ONE-WAY ANALYSIS OF VARIANCE FOR 115 CASES
   DEPENDENT VARIABLE IS SALESAMT
   GROUPING VARIABLE IS  ZIPCODE

     GROUP        COUNT    RANK SUM

       1.000      16       729.000
       2.000       9       447.000
       3.000      21      1212.500
       4.000      14       916.000
       5.000      18       822.500
       6.000       9       460.000
       7.000      13       843.000
       8.000      15      1240.000

   KRUSKAL-WALLIS TEST STATISTIC =      15.080
   PROBABILITY IS          0.035 ASSUMING CHI-SQUARE DISTRIBUTION WITH  7 DF
```

We have $8 - 1 = 7$ df The observed value of the test statistic, 15.08, corresponds to a p-value of 0.035.

b. The Kruskal-Wallis test is slightly more conclusive. The F test assumes normality, and is less efficient in the presence of outliers. Because the data are outlier-prone, we could expect the Kruskal-Wallis test to be more effective. It was.

Review Exercises—Chapters 9–11

R.71   **a.** First, we need the pooled variance. Note that $n_1 = n_2 = 40$.

$$s_{pooled}^2 = \frac{(40-1)(14.7)^2 + (40-1)(17.8)^2}{40 + 40 - 2} = 266.465$$

$$s = \sqrt{266.465} = 16.32$$

For a 95% confidence interval, we need $t_{0.025}$ with $40 + 40 - 2 = 78$ df.
Conservatively, we could use $t_{0.025,\ 60\ df} = 2.000$. The confidence interval is

$$(41.15 - 45.55) - 2.000(16.32)\sqrt{\frac{1}{40} + \frac{1}{40}} \le \mu_1 - \mu_2 \le (41.15 - 45.55) + 2.000(16.32)\sqrt{\frac{1}{40} + \frac{1}{40}}$$

$$-11.70 \le \mu_1 - \mu_2 \le 2.90$$

b. H_0 must be retained at $\alpha = 0.05$ because $\mu_1 - \mu_2 = 0.00$ is included in the 95% confidence interval.

c. The data should be regarded as paired samples, because the same people rated both formulations. Pooled variance procedures are for independent samples, not paired samples.

R.72   **a.** The 95% confidence interval is shown under "`tinterval`" as

$$2.15 \le \mu_{\text{diff}} \le 6.65$$

b. Yes, we can reject the null hypothesis that the mean difference is 0; $\mu_d = 0.00$ is not included in the interval.

c. The test statistic is shown in the output as `T = 3.95`. Because $n = 40$, we should look in the t table (Appendix Table 4) for $40 - 1 = 39$ df. For a two-tailed α of 0.05, we want a one-tail area equal to 0.025. Whether we use the entry for 30 df (2.042) or for 40 df (2.021) or something in between, it doesn't matter; the calculated statistic is larger than the table value. Therefore, reject H_0, just as we did using the confidence interval.

    If we look at the p-value, we find it to be 0.0003. This number is less than $\alpha = 0.05$. Using the Universal Rejection Region, reject H_0, just as we did using the confidence interval.

d. The data appear close to normally distributed. Perhaps the 23 score is a mild outlier, but not a serious one. Assuming that the ratings differences were based on a random sample of raters, and that there was no carryover effect of one rater's opinion to another's, there is no obvious violation of assumptions.

R.73   The output shows a p-value of 0.001, clearly smaller than $\alpha = 0.05$. Therefore, reject H_0, just as did the t test.

R.74   In Exercise **R.71a**, the width of the interval is $2.90 - (-11.70) = 14.60$. In Exercise **R.72a**, the width is $6.65 - 2.15 = 4.50$, a much smaller number. Treating the data (correctly) as paired yields a much more precise confidence interval. Pairing, by having the same tasters rate both formulations, made the statistical comparison much more effective. It controlled for variation among raters in their typical scores.

R.75   **a.** The value of the F statistic is shown in the output as `F = 4.26`. It has 3 and 60 df. The value in Appendix Table 6 for $\alpha = 0.01$ and these df is 4.13. Because the calculated F is greater than the tabled value, reject H_0. There is rather strong evidence that at least some of the population means differ. The apparent differences among sample means are unlikely to have occurred by chance.

b. Having rejected H_0 at $\alpha = 0.01$, we know that p-value < 0.01. The F table value for $a = 0.005$ is 4.73, so H_0 would be retained at $\alpha = 0.005$. Thus

$$0.005 < p\text{-value} < 0.01$$

In fact, the *p*-value is shown in the output as 0.0087.

c. We have no indication of any bias, though it may be present. The sample sizes are equal, so the difference of sample variances shouldn't be a problem. There is no reason to suspect dependence. To check normality, we can construct stem-and-leaf displays for each sample.

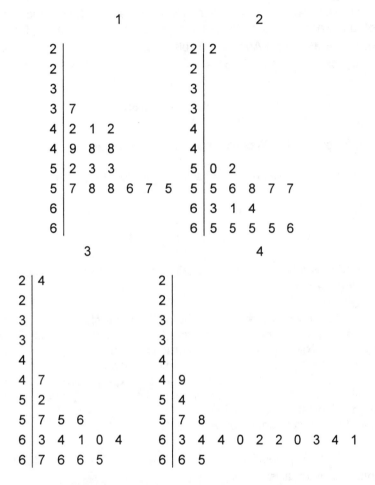

Obviously there are two extreme outliers. Perhaps there is left-skewness as well, but not severely so. The outliers strongly suggest nonnormal populations. Typically, outliers make mean-type tests (such as *F*) conservative, if anything. Thus our rejection of H_0 is not invalidated. A test that doesn't assume normality might be even more conclusive.

R.85   The output shows MS(Program) = 12.1 and MS(Error) = 25.5. Therefore $F = 12.1/25.5 = 0.47$. Because this value is less than 1 (the approximate expected value of F if H_0 is true), it can't be close to statistically significant. Appendix Table 6 does not have entries for 2 and 54 df; interpolating very roughly between the entries for 2, 40 and 2, 60 gives $F_{0.05, 2, 54} \approx 3.2$. Because F is so much less than 3.2, we must retain H_0 and declare the results not significant (detectable).

R.86   To compare row (program) means for this balanced design, we should use the Tukey method. Each row mean is the average of $2(10) = 20$ scores. MS(Error) is 25.5, as in Exercise **R.85**. From Appendix Table 8, $q_{0.05}(3.54) \approx 3.41$. The Tukey ± for a 95% confidence interval ($\alpha = 0.05$) is

$$3.41\sqrt{\frac{25.5}{20}} = 3.85$$

We can construct 95% confidence intervals such as

$$(36.50 - 35.80) - 3.85 \leq \mu_1 - \mu_2 \leq (36.50 - 35.80) + 3.85$$

or

$$-3.15 \leq \mu_1 - \mu_2 \leq 4.55$$

All these intervals include 0 so no differences of sample means are less than 3.85 so no differences are significant.

R.87   For inexperienced preparers (n in the plot), the profile (1 to 2 to 3) decreases, then increases. For experienced preparers (y in the plot), the profile increases, then decreases. There's an apparent interaction. The program means allow equal weights to experienced and inexperienced preparers, contradicting the experience of the service. Program 1 appears best for the experienced preparers; it clearly has the lowest mean of the three for experienced preparers. Thus the program means are not good indicators. In passing, note that the experienced preparers seem to take longer than inexperienced ones! Old dogs, new tricks?

R.93   a.   The sample proportions are $\hat{\pi}_1 = 36/100 = 0.36$ and $\hat{\pi}_2 = 48/100 = 0.48$. The 95% confidence interval is

$$(\hat{\pi}_1 - \hat{\pi}_2) - z_{0.025}\sqrt{\frac{\hat{\pi}_1(1-\hat{\pi}_1)}{n_1} + \frac{\hat{\pi}_2(1-\hat{\pi}_2)}{n_2}} \le \pi_1 - \pi_2 \le (\hat{\pi}_1 - \hat{\pi}_2) + z_{0.025}\sqrt{\frac{\hat{\pi}_1(1-\hat{\pi}_1)}{n_1} + \frac{\hat{\pi}_2(1-\hat{\pi}_2)}{n_2}}$$

$$(0.36 - 0.48) - 1.96\sqrt{\frac{0.36(0.64)}{100} + \frac{0.48(0.52)}{100}} \le \pi_1 - \pi_2$$

$$\le (0.36 - 0.48) + 1.96\sqrt{\frac{0.36(0.64)}{100} + \frac{0.48(0.52)}{100}}$$

$$-0.256 \le \pi_1 - \pi_2 \le 0.016$$

b. Because 0 is included in the interval, we must retain H_0: $\pi_1 - \pi_2 = 0$.

R.94 H_0: $\pi_1 - \pi_2 = 0$

$$z = \frac{(0.36 - 0.48) - 0}{\sqrt{\frac{0.36(0.64)}{100} + \frac{0.48(0.52)}{100}}} = -173$$

at $\alpha = 0.05$ (two-tailed) we would reject H_0 if $|z| > 1.96$. It is not; we don't have enough evidence to claim that there is a difference.

R.95   A basic assumption underlying the binomial distribution is that trials (days, here) are independent. If there are carryover effects, there is dependence from one trial to the next, violating independence.

R.96   The output shows a value of $\chi^2 = 9.29$ with 3 df. The p-value is shown as 0.0257. We have a fair amount of evidence that there is a more-than-random relation in the data.

R.97   The key assumptions are independence (the problem cited in Exercise **R.95**), constant probability over trials, and all expected frequencies greater than 5. There's no reason to suspect a trend in probabilities, unless the service is, say, gradually improving. Among expected frequencies, the smallest one will be in the row and column with the smallest frequencies. The rows are equal-sized, at 100 each. The "Poor" column has the lowest frequency. The expected frequency for (Current, Poor) is $(100)(12)/200 = 6.0$, as is the expected frequency for (Computerized, Poor). Thus there is no reason for concern about too-small expected frequencies. The warning in the output is misleading, because the issue is the expected counts not the observed counts in the table. There are no evident additional violations.

Chapter 12

Linear Regression and Correlation Methods

12.1   The Linear Regression Model
12.2   Estimating Model Parameters

12.1   **a.**   The line on the plot gives the predicted values. We need to read the predicted Y value off the vertical axis. When $X = 1$, the predicted value is somewhere between 13 and 17, closer to 17. We'd say it's about 16. When $X = 7$, on the right of the plot, the predicted value is just about 25. Therefore, a rough guess for the slope is (difference in predicted Y values) – (difference in X values) $= (25 - 16)/6 = 15$.

b.   Again, we want to read the predicted Y value along the vertical axis. When LOGX equals 0.0, the predicted Y seems to be halfway between 13 and 17, so call it 15. At the right side of the plot, the predicted Y is somewhere between 21 and 25, a little closer to 25; call it 24. At the right side the LOGX value seems to be a little less than 0.9, but close enough to call it 0.9. So a rough approximation for the slope is $(24 - 15)/(0.9 - 0.0) = 10$.

c.   The plot of Y vs. LOGX appears more nearly linear to us. In the plot of Y vs. X, the points in the middle tend to be above the line while the points at both ends tend to be below. In the plot of Y vs. LOGX, the points seem to be more evenly distributed above and below the line, all along that line.

12.2   **a.**   The first part of the output shows a regression equation for y as dependent variable and x as independent variable. The equation is shown as $y = 14.3 + 1.48x$. The coefficients are shown to more decimal places in the "Coef" column, as 14.2917 for the intercept (constant) and 1.4750 for the slope (coefficient of x). Notice that our "eyeball" guess for the slope, 1.5, was reasonable. And no, we didn't peek before making the guess.

b.   The residual standard deviation is shown in the output as $s = 1.346$. We can check this answer by taking the square root of MS(Residual). That mean square is shown in the output as MS(Error) = 1.81. By calculator, the square root of 1.81 is 1.345, equal to s up to roundoff error.

12.3 **a.** The second part of the Minitab output shows a regression of y vs. log x. The equation is shown as $y = 14.9 + 10.5\log(x)$. In the "Coef" column, the values to more places are 14.8755 and 10.522.

b. The residual standard deviation is $s = 1.131$. As a check, MS(Error) is 1.28, and the square root of 1.28 turns out to be 1.131.

12.4 The residual standard deviation in Exercise **12.3** is smaller. This result concurs with the choice of model based on the plot of Exercise **12.1**. We thought that the LOGX model fit the data better; one consequence of fitting the data better is more accurate prediction, hence a smaller standard deviation.

12.14 **a.** The LOWESS smooth is basically a straight line, with a slight bend at the high end. We would call that a linear relation.

b. The points with high leverage are those that are far from average along the x axis. In this case there are two points with very high Income values. They are the last two in the data list. The plot shows very clearly that the point with Income equal to 65.0 and Price equal to 110.0 is well away from the line. This point therefore has high influence. The other high-leverage point falls reasonably close to the line, and would not have such great influence.

12.15 **a.** The overall equation, shown first, has the coefficients of the line in the "Coef." column. The slope of Income is 1.80264, and the intercept is 47.15048.

b. The slope, 1.80, may be interpreted as a comparison of two purchasers. The purchaser who has 1 (thousand dollars per year) higher income is predicted to spend 1.80 (thousand dollars) more for a house. The intercept term would be the predicted price of a house bought by someone with no income. That seems a rather unlikely possibility; no one in the data had an income close to 0. Therefore, the intercept isn't directly meaningful.

c. The residual standard deviation is shown in the output as Root MSE = 14.445. As a check, it is also the square root of MS(Residual); the square root of 208.667027 is 14.445.

12.16 The slope changes quite a bit, from 1.80 to 2.46. The reason is that the omitted point is a high-influence outlier. Such a point twists the equation toward it, and can distort the slope a great deal.

12.3 Inferences about Regression Parameters

12.19 **a.** A plot of the data is shown below. Any computer package will do something similar. A rough approximation by hand should reveal something like this pattern.

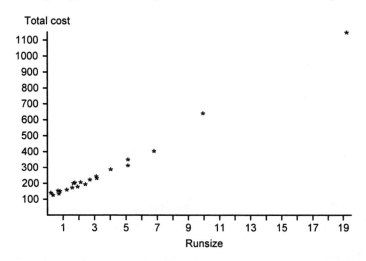

Examining the plot, a linear equation relating y to x seems plausible. There is one extreme point at runsize $= 20$, total cost $= 1,146$. There are no evident violations of constant variance or indenpendence assumptions.

b. $\hat{\beta}_1$ is in the Estimate column for Slope in the output: 51.9179

$\hat{\beta}_0$ is in the Estimate column for Intercept in the output: 99.777

Therefore, the estimated regression equation is

$$\hat{y} = 99.77704 + 51.9179x$$

The residual standard deviation is labelled Standard error of estimation in the output:

$$s_\varepsilon = 12.2065$$

c. The 95% confidence interval for β_1 is

$$\hat{\beta}_1 - t_{\alpha/2}s_{\hat{\beta}_1} \leq \beta_1 \leq \hat{\beta}_1 + t_{\alpha/2}s_{\hat{\beta}_1}$$

where

$$\alpha = 0.05 \text{ and } df = 28 \text{ so that } t_{\alpha/2} = 2.048$$

The output shows the slope as

$$\hat{\beta}_1 = 51.9179$$

and the standard error as

$$s_{\hat{\beta}_1} = 0.586455$$

Therefore,

$$51.9179 - (2.048)(0.586455) \leq \beta_1 \leq 51.9179 + (2.048)(0.586455)$$
$$50.72 \leq \beta_1 \leq 53.12$$

The slope, β_1, could be interpreted as the variable cost per sticker, while the intercept, β_0, could be interpreted as the fixed cost.

12.20 **a.** The value of the t statistic is located under the label "t Value" on the computer output:

$$t = 88.53.$$

b. The p-value is located under the label "P Value": p-value = 0.0000, and is therefore some number less than 0.00005. This p-value is a two-tailed p-value, because no direction for a research hypothesis is shown in the output. In principle, we probably should use a one-sided research hypothesis. Nobody would argue that the total cost of a job should go down as it gets larger. However, dividing 0.0000 by 2 still gives 0.0000.

12.21 **a.** The F value is located under the label "F-Ratio" on the output: $F = 7,837.26$. The associated p-value is located under the label "P Value": p-value = 0.0000.

b. The conclusions of the F test and the t test are identical, since in simple regression models $F \approx t^2$.

12.4 Predicting New Y Values Using Regression

12.22 **a.** The regression equation is

$$\hat{y} = 99.77704 + 51.9179x$$

Therefore, if $x_{n+1} = 2.0$, then

$$E(Y_{n+1}) = \hat{y}_{n+1} = 99.77704 + 51.9179(2.0) = 203.613$$

as shown in the output.

b. The 95% confidence interval for the mean is shown in the output under "Confidence Limits." The "Prediction Limits" heading is for an individual value. One way to check this is to note that the individual value limits must be wider that the limits for a mean. The output indicates a 95% confidence interval for the mean as

$$198.902 \le E(Y_{n+1}) \le 208.323$$

12.23 $x_{n+1} = 2.0$ is close to the mean of the x values used in determining the prediction equation. Therefore, we would not expect major extrapolation in the prediction.

12.24 **a.** The predicted value will be exactly the same as the predicted value for the mean, 203.613. The interval for an individual value is what we want now. We're trying to predict one particular value, not an average. The desired interval is shown under "Prediction Limits." The limits are

$$178.169 \le Y_{n+1} \le 229.057$$

b. Yes, $250 does not fall in the 95% prediction interval. It is quite a bit higher than even the upper limit, 229.057.

12.5 Correlation

12.29   The output shows `R-square` = `0.9452`. The slope of the regression equation is positive, indicating an increasing relation. Therefore, the correlation will also be positive. The correlation is the positive square root of 0.9452, namely 0.9722.

12.30   **a.** The hypothesis test for ρ_{yx} is based on a t statistic. The standard five-step procedure is the following; we use a one-sided research hypothesis, as we did in testing the slope for this situation.

H_0: $\rho_{yx} = 0$

H_a: $\rho_{yx} > 0$

T.S.: $t = \dfrac{r_{yx}\sqrt{n-2}}{\sqrt{1-r_{yx}^2}} = \dfrac{0.9722326\sqrt{12-2}}{\sqrt{1-(0.9722326)^2}} = 13.138$

R.R.:   For $\alpha = 0.05$ with df $= 10$, reject H_0 if $t > 1.812$

Conclusion: Reject H_0.

b. The t test of the slope, shown in the output, is algebraically identical to this test of correlation. We obtain $t = 13.138$ for both tests, as we must if we've done the tests correctly.

12.32   **a.** The correlation is shown as 0.956. This value does not have a direct numerical interpretation itself. We must square it to get an error reduction number. The squared correlation coefficient is $(0.956)^2 = 0.914$. Variation in intensity accounts for 91.4% of the variability in awareness, for these data.

b. We used a computer program to obtain the following plot of y vs. x.

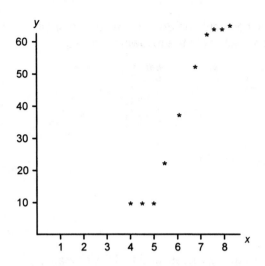

The relation between y and x doesn't appear linear. It does appear to be generally increasing. There seems to be a threshhold effect; at low values of x, the y value stays small. At larger values of x, the y values increase sharply. But then the values approach their upper limit, and level off again.

Supplementary Exercises

12.37 **a.** With only 7 points, one could plot the data either by hand or by computer. A plot y vs. x is shown below.

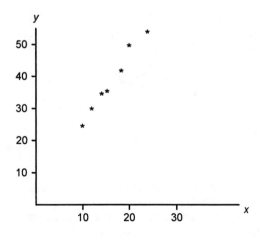

b. This calculation could be done by hand. As an alternative, we used the Excel spreadsheet program, with the following results.

x	y	x-mean(x)	y-mean(y)	product	(x-mean(x))sq
10	25	-5.8571	-14.2857	83.6735	34.3061
12	30	-3.8571	-9.2857	35.8163	14.8776
14	36	-1.8571	-3.2857	6.1020	3.4490
15	37	-0.8571	-2.2857	1.9592	0.7347
18	42	2.1429	2.7143	5.8163	4.5918
19	50	3.1429	10.7143	33.6735	9.8776
23	55	7.1429	15.7143	112.2449	51.0204

Mean	15.8571 39.2857				
Sum				279.2857	118.8571

Slope	2.3498
Intercept	2.0252

The steps were:

1. Enter the x and y values, and find their means.
2. Subtract the appropriate mean from each value and store in new columns.
3. Multiply the deviations and store them in the "product" column; sum the products.
4. Square the x deviations, store them, and sum them.
5. Divide the sum of products by the sum of squared x deviations; that is the slope.
6. Take the y mean minus the slope times the x mean; that is the intercept.

The slope is shown as 2.3498, and the intercept as 2.0252.

c. When $x_{n+1} = 21$, the predicted value of y is

$$\hat{y} = 2.0252 + 2.3498(21) = 51.37$$

12.38 a. Again, we used Excel to do the arithmetic.

x	y	predicted	residual	residual squared
10	25	25.5228	-0.5228	0.2734
12	30	30.2224	-0.2224	0.0494
14	36	34.9219	1.0781	1.1624
15	37	37.2716	-0.2716	0.0738
18	42	44.3209	-2.3209	5.3866
19	50	46.6707	3.3293	11.0844
23	55	56.0697	-1.0697	1.1443

Mean	15.8571 39.2857			
Sum				19.1743
			st.dev.	1.9583

The steps were:

1. Create and store predicted values by multiplying each x by the slope and adding the intercept.
2. Subtract the predicted value from the actual y value and store these residuals.
3. Square the residuals and sum them.

4. Divide by $7 - 2 = 5$, and take the square root. That is the residual standard deviation, 1.9583.

b. The residuals were calculated in the preceding answer. The interval within $\pm 2s_\varepsilon$ of 0 is

$$-2(1.9583) \le (y - \hat{y}) \le 2(1.9583)$$
$$-3.9166 \le (y - \hat{y}) \le 3.9166$$

All of the residuals for these data lie within this interval.

12.44 **a.** The output contains the regression coefficients in the COEFFICIENT column. The predicted value for gallons used is $140.074 + 0.61896$ Miles.

b. The coefficient of determination is another name for R^2. This value is shown as 0.9420 in the output. It means that variation in flight miles accounts for 94.20% of the variation in fuel use. The correlation coefficient must be the square root of R^2; because the slope is positive, the correlation must be positive. Therefore the correlation is $\sqrt{0.9420} = 0.9706$. The correlation number does not have a direct interpretation; we can interpret the squared correlation more effectively.

c. There is no point in testing $H_0: \beta_1 = 0$. Obviously, longer flights take more fuel; the slope β_1 must be positive.

12.45 **a.** The predicted y value when $x = 1{,}000$ can be found by evaluating the regression equation at $x = 1{,}000$. Predicted $y = 140.074 + 0.61896(1000) = 759.03$. This is exactly the number shown as PREDICTED VALUE and also as FITTED VALUE in the output. Therefore, it is clear that the program has been asked to carry out the prediction at 1,000 miles.

There are two sets of LOWER and UPPER values indicated in the output, just as there are two predictions (mean or individual value) that can be made. To see which set we should use, note that predicting the mean must always yield a narrower interval than predicting an individual value. For this part of the exercise, we are predicting the mean, and want the narrower range. The narrower lower and upper limits are indicated by FITTED BOUND. The output also indicates that PERCENT COVERAGE is 95.0, so these are 95% intervals. Therefore,

$$733.68 \le E(Y_{n+1}) \le 784.38$$

b. Now we are predicting an individual flight's fuel usage. The best estimate is the same as the estimate of the expected value, 759.03. We want the wider of the two indicated 95% intervals. This is the one shown in the PREDICTED BOUND part of the output. We can read the prediction interval as

$$678.33 \le Y_{n+1} \le 839.73$$

A value for *y* equal to 628 does not fall in this interval. It would be regarded as unusually low, and worth investigation to see why the flight did so well on usage.

12.46 The value of β_1 is the fuel used for each mile of air mileage.

   The value of β_0 might be interpreted as the fixed cost, the amount of fuel needed to take off (and land), over and above the fuel needed in flight.

12.52 **a.** The group felt that small "townpopn" should be associated with large "expendit" (and presumably that large "townpopn" should go with smaller "expendit"). Thus, the group claimed that there is an inverse (negative) relation between the two variables. The slope should have a minus sign, according to the group.

   b. The slope is shown as a *positive* number, 0.0005324. The slope is not negative, so it does not confirm the opinion of the group.

12.53 There is one point at the upper right of the plot that is far off any line that could reasonably be drawn through the other points. This high-influence outlier is twisting the line toward it. In this case, the twist tends to increase the slope greatly. The line seems very misleading, because of this point.

12.54 **a.** The point is a very high-influence outlier, which distorted the slope a great deal. The intercept also was distorted a great deal. Removing the outlier therefore had a large effect on the regression line.

   b. The new line goes down as population increases, indicating that the slope is negative. The negative slope does confirm the opinion of the group, which had argued that the smallest towns would have the highest per capita expenditures.

12.55 The slope with the unusual town included was 0.0005324. The output with the unusual town excluded is shown in the output as –0.0015766. The slope has changed sign, from positive to negative. As it happens, it has also gotten larger in magnitude.

12.59 **a.** We used Execustat to analyze these data. The correlation coefficient was obtained using the "Relate" portion of the menu.

Correlation Analysis

	Sales	Density
Sales		-0.7707
Density	-0.7707	

The table shows estimated product-moment correlation

The negative sign indicates that sales tend to decrease as density increases. This makes reasonably good sense; lawn-care companies tend to sell services to larger homes, in low-density neighborhoods.

b. Execustat regression output is shown here. Other packages should give similar results.

Simple Regression Analysis

Linear model: Sales = 141.525 - 12.8926*Density

Table of Estimates

	Estimate	Standard Error	t Value	P Value
Intercept	141.525	9.10863	15.54	0.0000
Slope	-12.8926	1.94595	-6.63	0.0000

R-squared = 59.40%
Correlation coeff. = -0.771
Standard error of estimation = 21.7406
Durbin-Watson statistic = 1.82147
Mean absolute error = 16.5815

The intercept term, 141.525, is the predicted sales for a ZIP code with 0 density (that is, lawn-care sales in a ZIP code area with no homes—hmmmm). The slope indicates that, comparing two ZIP code areas, an area with 1 extra home per acre is predicted to have sales that are 12.8926 lower.

c. In Execustat, and several other packages, the residual standard deviation is shown as the Standard error of estimation. In the output for this data set, it is shown as 21.7406. Alternatively, we may use the following ANOVA table.

Analysis of Variance

Source	Sum of Squares	DF	Mean Square	F-Ratio	P Value
Model	20747.2	1	20747.2	43.90	0.0000
Error	14179.6	30	472.654		
Total (corr.)	34926.9	31			

The residual standard deviation is the square root of Mean Square Error. By calculator, the square root of 472.654 is 21.7406, as we found previously. As a rough rule, 95% of the predicted sales will be accurate within ±2 standard deviations; here 2 standard deviations is 43.4812, so 95% of the predictions will have an error less than about 43.5.

12.60 **a.** In the output for part **b.** of the previous question, the t statistic is shown as -6.63, with a p-value of 0.0000. The very small p-value indicates very conclusive evidence that density has value in predicting sales.

b. The output from part **b.** of the previous question shows the slope as -12.8926 and the standard error as 1.94595. There are 32 observations in the data, so the df for error are $32 - (1+1) = 30$; these df are also shown in the ANOVA table of the previous exercise. The *t* table value for a one-tail area 0.025 and 30 df is 2.042. Therefore the desired confidence interval is $-12.8926 \pm 2.042(194595)$, or $-16.87 \leq \text{slope} \leq -8.92$.

12.61 A plot of the data, as produced by Statgraphics, is shown here. Other packages will yield a plot with slight differences in appearance but the same essential pattern.

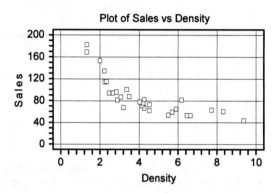

Plot of Sales vs Density

There is a definite curve in the plot. Therefore, straight-line prediction is not the best choice.

12.62 **a.** The density variable is number of homes per acre. Therefore, 1/density is number of acres per home. In other words, it is the typical size of a lot. If this variable is 0.5, then a typical home in this ZIP code area has a one-half acre lot.

b. A plot of sales against lot size, produced by Statgraphics, is shown here.

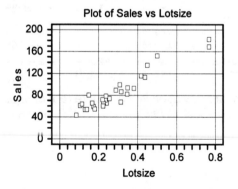

Plot of Sales vs Lotsize

There is no evident, pronounced curve in the data; the points seem to fall basically along a straight line. Therefore, a straight-line prediction seems quite reasonable.

c. Execustat yielded a correlation of sales and lot size.

```
                    Correlation Analysis

               Sales        Lotsize
Sales                       0.9517
Lotsize        0.9517

        The table shows estimated product-moment correlation
```

This correlation is larger in magnitude than the correlation of −0.7707 between sales and density. Correlation measures the extent of straight-line relation. The relation between sales and lot size is much more nearly linear than the relation between sales and density. Therefore, the correlation is stronger.

Chapter 13

Multiple Regression Methods

13.1   The Multiple Regression Model
13.2   Estimating Multiple Regression Coefficients

13.8   **a.** The estimated regression equation can be read from the `Estimate` column of the output.

$$\hat{y} = 50.0195 + 6.64357 \text{ Cat1} + 7.3145 \text{ Cat2} - 1.23142 \text{ Cat1Sq} - 0.7724 \text{ Cat1Cat2} - 1.1755 \text{ C}$$

b. The SS(Residual) is labeled `Error` and is located under the heading `Sum of Squares` on the computer output:

$$\text{SS(Residual)} = 71.489$$

The residual standard deviation is labeled `Standard error of estimation`:

$$s_\varepsilon = 2.25973$$

As a check, MS(Residual), called `Mean Square for Error` here, is `5.10636`. Its square root should be the residual standard deviation.

$$\sqrt{5.10636} = 2.25973$$

13.9   **a.** The R^2 value for the model is shown in the output as 86.24%, (0.8624). We can check this result by using the sums of squares.

$$R_{yx_1x_2x_3x_4x_5} = \frac{\text{SS(Total)} - \text{SS(Residual)}}{\text{SS(Total)}} = \frac{519.682 - 71.489}{519.682} = 0.8624$$

b. The conditional sums of squares for Cat1 and Cat2 are 286.439 and 19.3688, respectively. This fact means that the model containing only these two predictors would have SS(Model) $= 286.439 + 19.3688 = 305.8078$, and therefore SS(Error) $= 519.682 - 305.8078 = 216.8742$. The R^2 for the two-predictor model would be $305.8078/519.682 = 0.5885$.

13.3 Inferences in Multiple Regression

13.10 **a.** The F statistic is labeled F(3, 20) on the computer output:

$$F = 22.28$$

b. We can, if we wish, carry out the five steps of a hypothesis test.

H_0:	$\beta_1 = \beta_2 = \beta_3 = 0$
H_a:	At lease one $\beta_j \neq 0$
T.S.:	$F = 22.28$
R.R.:	At $\alpha = 0.01$ with $df_1 = 3$ and $df_2 = 20$, reject H_0 if $F > 4.94$

Conclusion: Reject H_0. The hypothesis of no overall predictive value can be rejected.

Alternatively, we could simply find the p-value, shown as Prob > F in the output. It is .0000, meaning some number smaller than 0.00005. The p-value is much smaller than $\alpha = 0.01$, so again we can reject H_0.

c. The t statistic for the coefficient ($\hat{\beta}_1$) of promotion is shown in the t column.

$$t = 4.842$$

d. Probably the easiest way to answer this question is to look at the p-value in the output, shown as P>|t|. The p-value for Promo is shown as 0.000. Whatever the exact number may be, it is smaller than 0.0005 and certainly smaller than $\alpha = 0.05$. Thus we can reject H_0 and support the stated research hypothesis.

Alternatively, we can carry out the five steps of a formal test.

H_0: $\beta_1 = 0$

H_a: $\beta_1 \neq 0$

T.S.: $t = 4.84$

R.R.:   At $\alpha = 0.05$, df $= 20$, reject H_0 if $|t| > 2.086$

Conclusion: Reject H_0

e. Reject H_0 in favor of H_a: $\beta_1 \neq 0$. That is, promotion has additional predictive value in predicting y, over and above that contributed by the other independent variables, development expenditure and reseach effort.

13.11 The p-value is located under the heading P>|T|:

$$p\text{-value} = 0.000$$

It is a two-tailed p-value, as indicated by the absolute value signs in the heading.

13.12 The results of the t tests are summarized below:

H_0	H_a	T.S. t	Conclusion
$\beta_0 = 0$	$\beta_0 \neq 0$	$t = 1351$	p-value $= 0.192$, retain H_0
$\beta_1 = 0$	$\beta_1 \neq 0$	$t = 4.842$	p-value $= 0.000$, reject H_0
$\beta_2 = 0$	$\beta_2 \neq 0$	$t = -1201$	p-value 0.244, retain H_0
$\beta_3 = 0$	$\beta_3 \neq 0$	$t = -0.904$	p-value $= 0.377$, retain H_0

Devel (development expenditure) and Research (research effort) each have not been proven to have additional predictive value in predicting y, as the "last predictor in." We have clear evidence that Promo (promotion) has additional predictive value in predicting y, as the "last predictor in."

13.4 Testing a Subset of the Regression Coefficients

13.20　**a.** The R^2 value for the reduced model (called Model 2 in the output) is labeled R-SQUARED:

$$R^2 = 0.2049$$

b. The complete model is Model 1 in the output, with all three predictors used. It shows an R^2 value of 0.7973.

c. The F statistic based on the incremental R^2 can be calculated from the given output.

$$F = \frac{\dfrac{R^2_{complete} - R^2_{reduced}}{k-g}}{\dfrac{1 - R^2_{complete}}{n - (k+1)}}$$

where

$$R^2_{complete} = 0.7973$$
$$R^2_{reduced} = 0.2049$$
$$k = \text{\# variables for complete model} = 3$$
$$g = \text{\# variables for reduced model} = 1$$
$$n - (k+1) = \text{df for error, complete model} = 17$$

Therefore,

$$F = \frac{\dfrac{0.7973 - 0.2049}{3-1}}{\dfrac{1 - 0.7973}{17}} = 24.84$$

The null hypothesis tested is

$$H_0: \beta_{BUSIN} = \beta_{COMPET} = 0$$

that is, the independent variables BUSIN and COMPET have no predictive value once INCOME is included as a predictor.

The computed value of F is much larger than any of the values in the F table for 2 and 17 df. Therefore, we reject the null hypothesis and conclude instead that at least one of BUSIN and COMPET adds predictive value, over and above that of INCOME.

13.21 Another way to compute the incremental F statistic is

$$F = \frac{\frac{SS(\text{Regression, complete}) - SS(\text{Regression, reduced})}{k - g}}{\frac{SS(\text{Residual, complete})}{n - (k+1)}} = \frac{\frac{2.65376 - 0.68192}{3 - 1}}{\frac{0.67461}{17}} = 24.84$$

13.5 Forecasting Using Multiple Regression

13.25 The prediction interval refers to predicting an individual value, as opposed to the confidence interval for the expected Y value. The output indicates that we want to Predict at x 3 w 1 v 6. These values are the new, $n + 1$ values. The output contains a 95% P.I.:

$$21.788 \le Y_{n+1} \le 44.212$$

The warning about very extreme X values may be surprising. Each new value is barely within the range of that independent variable. The value $x = 3$ is at the high end of that variable; $w = 1$ is at its low end; $v = 6$ is at the high end of the v data. In addition, the independent variables are collinear (correlated). In particular, w and v are positively correlated (correlation 0.262 from the output). Therefore, if one of these two variables is low, the other also tended to be low in the original data. We are asking to predict when w is low but v is high. This violation of the correlation structure, together with selection of values at the edge of the range of each variable, causes an extrapolation, which is the basis for the warning.

Supplementary Exercises

13.33   **a.** The multiple regression equation is

$$\hat{y} = -16.8198 + 147019\,X1 + 0.994778\,X2 - 0.0240071\,X1X2 - 0.01031\,X2SQ + 0.000249574\,X1\rangle$$

The residual standard deviation is labeled standard error of estimation on the computer output:

$$s_\varepsilon = 3.39011$$

b. The easiest way to test for "last predictor in" value is to use the t test shown in the output. The t statistic for $x_3 = x_1 x_2$ is shown as -1.01, with a p-value of 0.3243. This is not a statistically significant value at any reasonable α value. We don't have evidence to say that this variable adds predictive value, given all the others.

13.34   **a.** The complete model from the computer output of this exercise is just the one stated in the previous exercise. That model contained the larger number of predictors.

$$\hat{y} = -16.8198 + 147019\,X1 + 0.994778\,X2 - 0.0240071\,X1X2 - 0.01031\,X2SQ + 0.000249574\,X1\rangle$$

The reduced model from the computer output of this exercise is obtained from the Estimate column of the output.

$$\hat{y} = 0.010006 + 1.01603\,X1 + 0.0559262\,X2$$

b. To test whether a subset of the independent variables adds predictive value, given other variables, we need the F test based on incremental R^2 (or equivalently on the sequential sum of squares, not shown in this output).

H_0: $\quad\quad \beta_3 = \beta_4 = \beta_5 = 0$

H_a: $\quad\quad$ at least one of β_3, β_4, and β_5 is not 0

T.S.: $\quad\quad F = \dfrac{\frac{R^2_{complete} - R^2_{reduced}}{k-g}}{\frac{1-R^2_{complete}}{n-(k+1)}} = \dfrac{\frac{(0.9172-0.9064)}{5-2}}{\frac{1-0.9172}{26-(5+1)}} = 0.867$

R.R: $\quad\quad$ At $\alpha = 0.05$ with $df_1 = 3$ and $df_2 = 20$, reject H_0 if $F > 3.10$

Conclusion: Retain H_0. The improvement in R^2 obtained by adding x_3, x_4, and x_5 is not

statistically significant.

We note that we could not even reject the null hypothesis if we had used $\alpha = 0.25$; the tabled value is 1.48. Therefore, the p-value is even larger than 0.25.

$$p\text{-value} = P(F > 0.867) > 0.25$$

13.40 **a.** The regression equation is found from the `Coefficients` column of the output.

$$\hat{y} = 18.678 + 0.5420x_1 + 1.2074x_2 + 8.7779x_3 + 4.4211x_4 + 2.7165x_5 + 0.9225x_6$$

where x_1, x_2, x_3, x_4, x_5, and x_6 stand for the independent variables Senior, Sex, RankD1, RankD2, RankD3, and Doct, respectively.

b. An increase of 1 in the Sex variable necessarily is an increase from 0 to 1. The coefficient of Sex is the difference in mean base salary per year (in thousands of dollars) between males and females, for a given level of seniority, rank and holding of a doctorate. All else equal, a male professor is predicted to have a salary 1.2074 thousand dollars higher than a female professor.

c. The coefficient of RankD1 is the difference in mean base salary per year (in thousands of dollars) between full professors and lecturers, for a given level of seniority, sex and holding of a doctorate.

13.41 **a.** To test a single predictor, use the t test in the output. We could refer to the p-value, or we could carry out the steps of a test.

H_0: $\beta_2 = 0$

H_a: $\beta_2 > 0$

T.S.: $t = \dfrac{\hat{\beta}_2 - 0}{s_{\hat{\beta}_2}} = \dfrac{1.2074}{1.0649} = 1.134$

R.R.: At $\alpha = 0.05$ and df $= 23$, reject H_0 if $t > 1.714$

Conclusion: Retain H_0

This conclusion also follows because the p-value is large. Technically, we should divide the indicated, two-tailed p-value by 2, because we are doing a one-tailed test, and the sample coefficient does at least have the hypothesized sign. Even so, the p-value will still be larger than $\alpha = 0.05$.

b. From the test in part **a.**, H_0: $\beta_2 = 0$ is retained. Therefore, for a given level of all other factors, the mean base salary per year of men is not proved to be greater than that of women. The results don't conclusively prove a difference. Crucially, the test assumes that men and women have the same rank, doctorate, and seniority, because we are assuming that all other variables are given. It could be that discrimination occurs in these variables.

13.42 **a.** The value of the F statistic is shown as $F = 64.646$.

b. The null hypothesis to be tested by the F statistic in part **a.** is

$$H_0: \beta_1 = \beta_2 = \beta_3 = \beta_4 = \beta_5 = \beta_6 = 0$$

That is, we are testing that none of the independent variables, Senior, Sex, RankD1, RankD2, RankD3, and Doct, contributes to the prediction of salary. This is a rather silly hypothesis. Of course these variables (or at least some of them) matter to salary.

c. We could look at the very small p-value or carry out the steps of a test.

H_0: $\beta_1 = \beta_2 = \beta_3 = \beta_4 = \beta_5 = \beta_6 = 0$

H_a: at least one $\beta_j \neq 0$

T.S.: $F = 64.646$

R.R.: At $\alpha = 0.01$, $df_1 = 3$, and $df_2 = 26$, reject H_0 if $F > 3.71$

Conclusion: Reject H_0

p-value $= P(F > 64.646) \approx 0$. The rather absurd null hypothesis is very soundly rejected.

13.43   **a.** The R^2 for the reduced model is shown in the output as 0.9403.

b. We are testing two predictors together, so we should not use the one-by-one t tests. We need the F test based on incremental R^2 (or on incremental SS).

H_0:   the true coefficients of SEX and DOCT are 0

H_a: at least one coefficient is not 0

T.S.: $F = \dfrac{\dfrac{R^2_{complete} - R^2_{reduced}}{k-g}}{\dfrac{1 - R^2_{complete}}{n-(k+1)}} = \dfrac{\dfrac{(0.9440 - 0.9403)}{2}}{\dfrac{1 - 0.9440}{23}} = 0.760 < 1$ (Note: Under H_0,

$E(F) \approx 1$.)

R.R.:   At $\alpha = 0.01$ with $df_1 = 2$ and $df_2 = 23$, reject H_0 if $F > 5.66$

Conclusion: Retain H_0

We could also have used the sums of squares results.

13.44   **a.** The following output was obtained by using Execustat's multiple regression option within the "Relate" menu.

```
                Dependent variable: Salary

                                     Table of Estimates

                                        Standard              t
P
                        Estimate         Error             Value
Value

        Constant        25.5378        0.642972             39.72
0.0000
        Employees       0.00389372     0.00171822            2.27
0.0269
        Margin          0.0957243      0.0365267             2.62
0.0110
        IPCost          0.216348       0.0692004             3.13
0.0027
```

The equation predicts Salary to be

$25.5378 + 0.00389372$ Employees $+ 0.0957243$ Margin $+ 0.216348$ IPCost.

The coefficient of Employees indicates that, comparing two firms with equal margins and equal information processing costs, the firm with one more employee is predicted to pay an extra 0.0039 (thousand dollars, presumably) in salary. Similarly, comparing two firms with equal numbers of employees and equal IPCosts, the firm with a one-point higher margin is predicted to pay 0.096 thousand dollars more. A similar interpretation holds for the coefficient of IPCost.

b. To test the value of all predictors together, we need the following ANOVA table from Execustat.

Analysis of Variance

Source	Sum of Squares	Df	Mean Square	F-Ratio	P Value
Model	39.291	3	13.097	13.10	0.0000
Error	62.9831	63	0.999731		
Total (corr.)	102.274	66			

The F statistic is shown as 13.10, with a very small p-value, 0.0000. The data presents overwhelming evidence that the predictors jointly have at least some value in predicting Salary.

c. The "last predictor in" question refers to t tests, as shown in part **a.** All three predictors have p-values smaller than $\alpha = 0.05$. Therefore, we can reject each null hypothesis that the specific predictor adds nothing as "last predictor in." All three variables have detectable predictive value.

13.45 **a.** The Execustat output for this model also included the following results.

```
R-squared = 38.42%
Adjusted R-squared = 35.48%
Standard error of estimation = 0.999866
Durbin-Watson statistic = 2.01056
Mean absolute error = 0.764532
```

Converting from a percentage to a proportion, we find $R^2 = 0.3842$.

b. The Execustat output for the model with Employees as the only predictor follows.

Dependent variable: Salary

Table of Estimates

	Estimate	Standard Error	t Value	P Value
Constant	29,0841	0.234842	123.85	0.0000
Employees	0.00325929	0.00209794	1.55	0.1251

```
R-squared =  3.58%
Adjusted R-squared =  2.10%
Standard error of estimation = 1.23171
Durbin-Watson statistic = 1.87579
Mean absolute error = 0.983829
```

The R^2 value has decreased sharply to 0.0358.

c. The complete model has $R^2 = 0.3842$, based on $k = 3$ predictors. From the previous exercise, we have that $n - (k + 1) = 67 - (3 + 1) = 63$ df. The reduced model has $g = 1$ predictor. The incremental F test for the statistical significance of a subset of the predictors is therefore

$$F = \frac{\frac{0.3842 - 0.0358}{3 - 1}}{\frac{1 - 0.3842}{63}} = 17.82$$

The computed F value is far beyond all table values for 2 and 60 df, where we used 60 df in place of 63 df, which is not in our tables. There is overwhelming evidence that at least one of the omitted predictors would add predictive value.

13.46 The following table of correlations was obtained from Execustat. Virtually any program will provide a similar table.

	Salary	Employees	Margin	IPCost
Salary		0.1892	0.5045	0.5069
Employees	0.1892		0.0088	-0.1074
Margin	0.5045	0.0088		0.5315
IPCost	0.5069	-0.1074	0.5315	

The table shows estimated product-moment correlation

The only correlation between independent variables that is at all large is the 0.5315 correlation between Margin and IPCost. Even this value is not too large. Thus we wouldn't say that collinearity was a serious problem.

Chapter 14

Constructing a Multiple Regression Model

14.1 Selecting Possible Independent Variables (Step 1)
14.2 Using Qualitative Predictors: Dummy Variables (Step 1)
14.3 Lagged Predictor Variables (Step 1)

14.4   **a.** The dummy variables are Div2 and Div3. They are indicators of observations where the data come from division 2 or division 3, respectively. We need two dummy variables for the three categories (divisions).
That is, define

$$Div2 = \begin{cases} 1 & \text{if division} = 2 \\ 0 & \text{otherwise} \end{cases}$$

Define

$$Div3 = \begin{cases} 1 & \text{if division} = 3 \\ 0 & \text{otherwise} \end{cases}$$

If both Div2 = 0 and Div3 = 0, it follows by elimination that division = 1.

b. The interpretations assigned to the coefficients of these dummy variables can be seen by examining the corresponding expectations $E(Y)$. Plug in the appropriate 1's and 0's for the dummy variables.
The multiple regression equation is of the form

$$\text{predicted } Y = -1.8631 + 1.3542(\text{Forecast}) - 4.0066(\text{Div2}) + 0.9158(\text{Div3})$$

For division 1 (Div2 = 0, Div3 = 0):

$$E(Y) = -1.8631 + 1.3542(\text{Forecast})$$

For division 2 (Div2 = 1, Div3 = 0):

$$E(Y) = -1.8631 + 1.3542(\text{Forecast}) - 4.0066$$

For division 3 (Div2 = 0, Div3 = 1):

$$E(Y) = -1.8631 + 1.3542(\text{Forecast}) + 0.9158$$

It follows that –4.0066 is the difference in expected sales between division 1 and division 2 for a fixed forecast. Similarly, 0.9158 is the difference in expected sales between division 1 and division 3 for a fixed forecast.

c. The results of the *t* tests for Div2 and Div3 can be found under the headings t Stat and p-value on the computer output.

For Div2, the *t* statistic equals –5.67; the *p*-value equals 0.0000 (that is, some number less than 0.00005). Therefore, we reject H_0 that $\beta_2 = 0$ and conclude that Div2 has additional predictive value over and above that contributed by the other independent variables.

For Div3, the *t* statistic equals 1.25; the *p*-value equals 0.2205. Therefore, we do not reject H_0 that $\beta_3 = 0$ and assert that Div3 has not yet been shown to have additional predictive value over and above that contributed by the other independent variables. We can't yet say that division 3 differs from the baseline category, division 1.

14.5 We want to compare a regression model including the dummies to one excluding them. Therefore, we need a model that predicts actual sales without using the dummy variables. The following procedure would be used to test the null hypothesis that the dummy variables collectively have no predictive value, once the forecast value has been included in the regression equation.

H_0: $\beta_{\text{Div2}} = \beta_{\text{Div3}} = 0$

H_a: at least one of β_{Div2} and β_{Div3} is nonzero

T.S.: $F = \dfrac{\dfrac{R^2_{\text{complete}} - R^2_{\text{reduced}}}{k-g}}{\dfrac{1 - R^2_{\text{complete}}}{n - (k+1)}}$

R.R.: $F \geq F_a$

where "complete" represents a regression model using all predictors, Forecast, Div2, and Div3, and "reduced" represents a regression model using only the predictors which do not appear in H_0, i.e., Forecast.

In order for us to complete the above test, we would have to run a simple regression model (using FORECAST as the only predictor variable) and obtain computer printout so that we can calculate $R^2_{reduced}$. Note that computer output for the complete model is given; we therefore simply calculate $R^2_{complete}$ as

$$\frac{SS(Model)}{SS(CorrectedTotal)} = \frac{894.74508876}{968.30709677} = 0.924030$$

14.6   **a.**   The F statistic is testing the null hypothesis,

$$H_0: \ \beta_1 = \beta_2 = \beta_3 = 0$$

which says that none of the variables in the multiple regression has any predictive value at all. If this hypothesis is true, the textbook publisher has an utterly useless sales forecasting system.

b.   The result of the F test can be located under the heading F on the computer output, $F = 109.468$. At $\alpha = 0.01$, with 3 and 27 df, $F_a = 4.60$. The p-value is shown as 3.178E–15; in other words, it has 14 zeroes after the decimal point, followed by a 3. That's a small number. Therefore, we reject

$$H_0: \ \beta_1 = \beta_2 = \beta_3 = 0$$

and conclude that there is *some* degree of predictive value for at least one of the independent variables. This is not a very interesting conclusion; big deal!

14.7   **a.**   If the forecast sales had been exactly correct on the average within each division, the regression equation would be

$$\hat{y} = Forecast$$

The coefficient of "forecast" equals 1.0000, and the intercept equals 0.0000.

b.   A 95% confidence interval for the Forecast coefficient, $\hat{\beta}_1$, is

$$\hat{\beta}_1 - t_{\alpha/2}s_{\hat{\beta}_1} \le \beta_1 \le \hat{\beta}_1 + t_{\alpha/2}s_{\hat{\beta}_1}$$

where

$$\alpha = 0.05 \qquad t_{\alpha/2, \ df} = t_{0.025, \ 27} = 2.052$$

$$\hat{\beta}_1 = 1.3542 \text{ and } s_{\hat{\beta}_1} = 0.0829 \text{ (from the computer printout)}$$

Therefore,

$$13542 - (2.052)(0.0829) \le \beta_1 \le 1.3542 + (2.052)(0.0829)$$
$$1.3542 - 0.1701 \le \beta_1 \le 1.3542 + 0.1701$$
$$1.18 \le \beta_1 \le 1.52$$

The confidence interval doesn't include the value of 1 specified in part **a.**

14.4   Nonlinear Regression Models (Step 2)

14.16   **a.** The table of correlation coefficients is given on the computer output. For each pair of independent variables, MELT, CHILL, REPEL, SPEED, and KNIFE, we have correlation coefficient values of 0. Therefore, there is no collinearity at all in the data.

b. A plot of CHILL vs. MELT is given below. There are multiple points at each value, so there are fewer points shown in the plot than there are in the data.

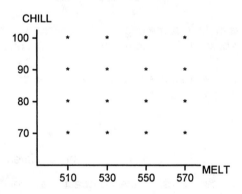

A plot of REPEL vs. MELT is given below:

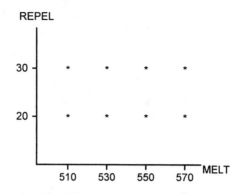

From the preceding two plots, plus the fact that there are equal numbers of observations at each of the points (check the data), we see a zero degree of correlation between MELT and CHILL, as well as MELT and REPEL. The points form a grid, so there is no way to predict the value of one variable given another. The values are perfectly balanced.

14.17 In each of the plots, there's no *obvious* evidence of nonlinearity. Perhaps the RESSTIFF vs. CHILL plot shows a slight curve (residuals at 80 and 90 are greater than at 70 and 100). It may be easier to look for where there are *no* points. There are "holes" in the plot of residuals vs. CHILL in the upper left and upper right corners. Still, there isn't any indication of severe nonlinearity.

14.18 **a.** The output indicates that the R^2 values are 0.904 and 0.914. To get a little more information, we can calculate based on the sums of squares. The R^2 for the second-order model is

$$R^2 = \frac{SS(\text{Regression})}{SS(\text{Total})} = \frac{3{,}141.625}{3{,}437.875} = 0.913828$$

The R^2 for the first-order model is

$$R^2 = 0.903581$$

The difference in R^2 values is

$$0.913828 - 0.903581 = 0.010247$$

b. We can carry out the five steps of the F test based on incremental R^2

H_0: $\beta_{MELTSQ} = \beta_{KNIFESQ} = 0$

H_a: at least one of β_{MELTSQ} and $\beta_{KNIFESQ}$ is non-zero

T.S.: $F = \dfrac{\dfrac{R^2_{complete} - R^2_{reduced}}{k-g}}{\dfrac{1 - R^2_{complete}}{n - (k+1)}} = \dfrac{\dfrac{0.913828 - 0.903581}{7 - 5}}{\dfrac{1 - 0.913828}{32 - (7+1)}} = 1.427$

R.R.:   At $\alpha = 0.05$ with $df_1 = 2$ and $df_2 = 24$, reject H_0 if $F > 3.4$

Conclusion: Retain H_0

We could also have based this test on the sums of squares.

c. Neither MELT*MELT (and MELT times MELT is just another way to write MELT squared) nor CHILL*CHILL is a statistically significant predictor as "last predictor in." We thought we might have seen some nonlinearity in the residual plot vs. CHILL, and the CHILL*CHILL term comes closer to significance; however, the apparent nonlinearity could well have occurred just by random variation. Note that for each of the t statistics, the associated p-value is greater than $\alpha = 0.05$, thereby indicating retention of H_0.

14.5   Choosing Among Regression Models (Step 3)

14.19   **a.** The order in which the independent variables enter the regression model can be read by looking at the successive steps. Alternatively, the variables are listed in the order entered, at the last step.

              REPEL, KNIFE, CHILL, MELT

b. From the correlation table of Exercise **14.16**, the independent variables in order of largest (in absolute value) to smallest correlation with STIFF are as follows:

              REPEL, KNIFE, CHILL, MELT, SPEED

c. The orderings of the variables given by the two lists in parts a and b are the same. The stepwise regression did not incorporate the SPEED variable at all, but it can fairly be thought of as the last variable in the list. The reason that the two lists are the same is the lack of any correlation among the independent variables. It doesn't matter what variables have been included previously, in this situation; the higher the correlation with STIFF, the better the predictor is, alone or in combination.

14.20 The following F test is used to test the hypothesis that the last two variables entered in the stepwise regression have no predictive value. Because the R^2 values are given only to three decimal places, there is a possibility of serious roundoff error. We did check the computations using other packages, and got the same conclusion, though somewhat different numerical values.

H_0: $\beta_3 = \beta_4 = 0$

H_a: at least one of β_3 and β_4 is not 0

T.S.: $F = \dfrac{\dfrac{R^2_{complete} - R^2_{reduced}}{k-g}}{\dfrac{1 - R^2_{complete}}{n-(k+1)}} = \dfrac{\dfrac{0.903 - 0.880}{4-2}}{\dfrac{1 - 0.903}{32 - (4+1)}} = 3.201$

R.R.:   At $\alpha = 0.05$ with $df_1 = 2$ and $df_2 = 27$, reject H_0 if $F > 3.35$

Conclusion: Do not reject H_0

Note: $R^2_{complete}$ complete is the R^2 value using all $k = 4$ predictors in the regression model (STEP 4 of the stepwise regression printout). $R^2_{reduced}$ is the R^2 value using only the $g = 2$ predictors that do not appear in H_0 (STEP 2 in the stepwise regression printout).

14.6   Residuals Analysis (Step 4)
14.7   Autocorrelation (Step 4)

14.24   **a.** The positive sign on the coefficient of Mileage indicates that, all else equal, as rated mileage increases, sales increase. The negative sign on the coefficient of GasPri indicates that as gas price increases (other variables constant), sales decrease. Similarly, an increase in PreGas, IntRate, or CarPrice (other things equal) predicts a decrease in sales, because each of these variables has a negative coefficient.

b. The residual standard deviation is labeled simply s and is equal to 47.54. We can also find the standard deviation by taking the square root of MS(Error).

$$s_\varepsilon = \sqrt{2{,}260} = 47.539$$

14.25 **a.** First, we should note that the data are time-series, so it would make sense to talk about the correlation of one residual with the next one. The Durbin-Watson statistic of 0.80 indicates that there is a problem of (positive) autocorrelation. Any value less than 1.5 or 1.6 leads us to suspect autocorrelation. In this case, the statistic is much lower than 1.5. So the auto sales data shows autocorrelation, sensibly enough.

b. Autocorrelation appears in residual plots vs time as a "snake-like," cyclical pattern. There is a rather strong cyclic pattern in the sequence plot of residuals. The first few residuals increase, then there is a clear decrease/increase/decrease pattern. There is a clear indication of autocorrelation.

14.26 **a.** The two variables that remain from the original model are interest rate, which has a negative sign but a smaller magnitude on the coefficient, and car price, which changed from a negative sign to a positive sign. However, the coefficient on the price variable is now very small and nowhere close to statistically significant.

Note that the Durbin-Watson statistic is now close to 2.0 and the sequence plot appears nearly random.

b. The residual standard deviation is labeled s and is equal to 37.34.

c. No, the residual standard deviation is smaller for the difference model than for the original model. Usually, a difference model is used to eliminate autocorrelation. When autocorrelation is present, the least squares prediction model provides too good a fit to the sample data. Because of this, the residuals are smaller than the true errors and the residual standard deviation, s_ε, is biased and provides an underestimation of the population standard deviation. Therefore, when autocorrelation is eliminated, we expect to see an increase in the residual standard deviation.

Supplementary Exercises

14.37 **a.** Lagged variables are measured at a time previous to the measurement of the dependent variable. The lagged variables in this exercise are:

1. Number of installations in previous month
2. Number of installations in 4 preceding months

b. Interaction is an "it depends" concept. Two predictors interact in their predicted effect on a dependent variable if the effect of changing one predictor depends on the level of the other. In the description, there was no indication of an "it depends" effect. We don't see why the effect of, say, an increase in this month's installations should depend on whether previous months were high, low, or in between. Therefore, there is no obvious indication that any severe interactions could be expected.

c. There is an indication of possible need for nonlinear terms in the regression model. There was some question as to whether support costs would increase in proportion to the number of installations. If cost per installation decreases as installations increase, we'd see the support cost increasing with installations, but at a decreasing rate. In other words, we'd expect a scatterplot of costs vs. number of installations to flatten out as the number of installations grew.

14.38 **a.** Yes, there are two dramatic outliers. These outliers are located at the bottom of the plot, at about months 16 and 17. They are far more negative than any other values.

b. Autocorrelation yields a cyclic, "snake-like" pattern in the sequence plot of residuals. In this plot, there appears to be an autocorrelation problem. A cyclical pattern can be seen from the plot. It is clearest in the earliest months, but we think we see some pattern later. The outliers are the most obvious problem, however.

14.39 Nonconstant variance yields a clear fan shape in a plot of residuals against predicted values. The pattern of residuals may indicate some pattern of increasing variance. The plot appears to widen somewhat as the predicted values increase. Therefore, there may well be a nonconstant variance problem.

14.40 **a.** The sequence has to be read by looking at the individual steps. Check at each step to see which variable has been added. The following list gives the sequence in which the independent variables are added in the stepwise regression analysis:

BCURRSQ, ACURRSQ, BPREV, APREV, BPREC4, BCURR, ACURR, APREC4

This is a strange sequence, with squared terms appearing before linear terms and lagged variables preceding current values.

b. The R^2 value for the regression model using only the first 4 variables is 0.9454. This R^2 value is located in the second column from the right of the regression analysis computer printout and labeled R SQ. The R^2 value for the regression model using all 8 of the independent variables is 0.9837.

Therefore, the increment to R^2 obtained by inclusion of the last 4 variables is $0.9837 - 0.9454 = 0.0383$.

14.41 **a.** The plot of standardized residuals shows a cyclical pattern. Therefore, there is evidence of autocorrelation. (The Durbin-Watson statistic (not shown) equals 1.34. Any value less than 1.5 or 1.6 leads us to suspect autocorrelation.)

b. Nonconstant variance is usually reflected in a fan-shaped plot of residuals against predicted values. Yes; in the plot of residuals vs. fitted (predicted) values, "the plot thickens" to the right. This pattern may not be obvious; notice that all the points in the left half of the plot are close to the axis, while several of the points in the right half of the plot are larger in magnitude.

14.42 Again, we must go step by step to see which variable is added at each step. The sequence in which the variables are added in the stepwise regression run (using differenced data) is as follows:

DACURR, DAPREV, DBCURR, DBPREV, DBCURRSQ, DAPREC4, DACURRSQ, DBPREC4

This sequence is different from the sequence in which the variables are added in the stepwise run (using the original data) in Exercise **14.40**.

14.43 **a.** The plot of the standardized residuals versus time (MONTH) indicates that there is no particular autocorrelation problem with the differenced data. The plot shows no obvious cyclical pattern.

b. There is no major indication of nonconstant variance in the plot. Perhaps the residuals are "fanning out" slightly, but we wouldn't say it was a severe problem.

14.44 **a.** The regression model is found in the RESULTING STEPWISE MODEL portion of the output, in the COEFFICIENT column.

$$\hat{y} = 0.52063 + 1.33473(\text{DACURR}) + 1.06333(\text{DAPREV}) + 0.119333(\text{DAPREC4})$$
$$+ 2.02548(\text{DBCURR}) + 0.39634(\text{DBPREV}) + 0.05833(\text{DBPREC4})$$
$$+ 0.00966(\text{DACURRSQ}) - 0.02003(\text{DBCURRSQ})$$

b. The output does not show an overall F statistic directly. We can calculate it from the R^2 value shown. We'll carry out the five steps, for clarity.

H_0: $\qquad \beta_1 = \beta_2 = \cdots = \beta_3 = 0$

H_a: $\qquad$ at least one of $\beta_1 \cdots \beta_8$ is not 0

T.S.: $\qquad F = \dfrac{\frac{R^2}{k}}{\frac{1-R^2}{n-(k+1)}} = \dfrac{\frac{0.9413}{8}}{\frac{1-0.9413}{29-(8+1)}} = 40.09$

R.R.: $\qquad$ at $\alpha = 0.05$ with 8 and 20 df, reject H_0 if $F > 2.45$

Conclusion: Reject H_0 that all partial slopes are 0, at $\alpha = 0.05$ (and at all other tabled values)

c. The t test is used to determine which variables have coefficients significantly different from 0 as "last predictor in":

H_0: $\beta_i = 0$ $i = 1, \cdots, 8$

H_a: $\beta_i \neq 0$

T.S.: t as shown on computer output

R.R.:   At $\alpha = 0.05$, reject H_0 if $|t| \geq 2.086$ (t table)

Conclusion: From the computer output, we see that 4 variables have associated t test statistic values greater than or equal to 2.086. Therefore, DACURR, DAPREV, DBCURR, and DBPREV are variables whose coefficients are significantly different from 0 at the $\alpha = 0.05$ level. Equivalently, these 4 variables have p-values smaller than $\alpha = 0.05$. We cannot conclude that the variables DBCURRSQ, DAPREC4, DACURRSQ, and DBPREC4 have coefficients that are significantly different from 0.

d. The R^2 value is labeled R SQUARED on the printout and is equal to 0.9413. Thus, 94.13% of the variability of the y value is accounted for by variability in DACCUR, DAPREV, DBCURR, DBPREV, DBCURRSQ, DAPREC4, DACURRSQ, and DBPREC4.

e. The residual standard deviation is labeled SD on the output and is equal to 3.35551.

14.45 **a.** The residuals are simply the difference between the actual and predicted support levels, and are calculated below.

Actual Support	Predicted Support	Residual	=	Actual Support	−	Predicted Support
2	5.78			−3.78		
−3	−0.97			−2.03		
5	2.99			2.01		
−5	−9.94			4.94		
17	15.29			1.71		
11	12.78			−1.78		

The standard deviation of the residuals is 3.2482, by calculator.

b. The residual standard deviation found in Exercise **14.42** was 3.35551. The residual standard deviation in part **a.** of this exercise is actually a bit smaller, but essentially similar.

c. Yes, the regression model appears to be useful in predicting the budget figure. If we allow ± 2 standard deviations for prediction error, that's $2(3.25) < 7$ units compared to ± 20.

14.59 **a.** We used Systat to analyze these data. The results from other packages will be very similar.

```
PEARSON CORRELATION MATRIX

                        SQFT      SKILLED      HELPERS     VEHICLES

        SQFT           1.000
        SKILLED        0.888      1.000
        HELPERS        0.837      0.819        1.000
        VEHICLES       0.853      0.862        0.805       1.000

NUMBER OF OBSERVATIONS:    36
```

There is a serious collinearity problem. All the predictor variables have correlations with each other that are larger than 0.8. Presumably, this is because all the independent variables basically reflect the size of the project. Therefore, it's natural that they tend to be large or small together.

b. None of the variables are qualitative or ordinal; all of them are actual measurements. Therefore, none of them need to be replaced by dummy variables. The data are not really time series. There is no point in predicting the progress on one job by the resources used in another, preceding job. Therefore, no lagged variables are needed.

c. Regression output from Systat's MGLH module are shown next.

```
DEP VAR:    SQFT    N:  36   MULTIPLE R:   .916    SQUARED MULTIPLE R:   .838
ADJUSTED SQUARED MULTIPLE R:   .823    STANDARD ERROR OF ESTIMATE:   11.199

VARIABLE    COEFFICIENT   STD ERROR    STD COEF TOLERANCE    T    P(2 TAIL)

CONSTANT       18.402        9.784     0.000 1.0000000     1.881    0.069
SKILLED         4.248        1.401     0.468  .2121126     3.032    0.005
HELPERS         2.493        1.255     0.262  .2910425     1.986    0.056
VEHICLES        6.644        4.159     0.239  .2266293     1.598    0.120

                        ANALYSIS OF VARIANCE

   SOURCE    SUM-OF-SQUARES    DF   MEAN-SQUARE     F-RATIO        P

 REGRESSION     20798.640      3     6932.880       55.278       0.000
   RESIDUAL      4013.360     32      125.418

WARNING: CASE    8 HAS UNDUE INFLUENCE (LEVERAGE =        .412)
WARNING: CASE    9 IS AN OUTLIER (STUDENTIZED RESIDUAL =       3.205)

DURBIN-WATSON D STATISTIC      2.296
FIRST ORDER AUTOCORRELATION   -.178
```

Predicted sqft equals 18.402 + 4.248 skilled + 2.493 helpers + 6.644 vehicles. The coefficient of determination is shown at the top of the output, as SQUARED MULTIPLE R = 0.838.

14.60 **a.** Systat provided a plot of residuals against predicted values, called estimates, as shown.

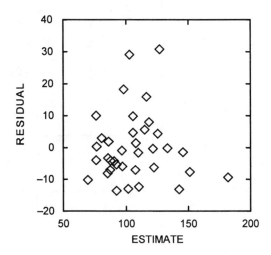

There may be some tendency for a wider spread around 0 for large predicted values, but it's not obvious or glaring to us. It does seem that many of the large residuals occur at relatively large predicted values. There are no other obvious problems; in particular, we don't see any outliers.

b. The Systat plots of residuals against the three predictors are shown here.

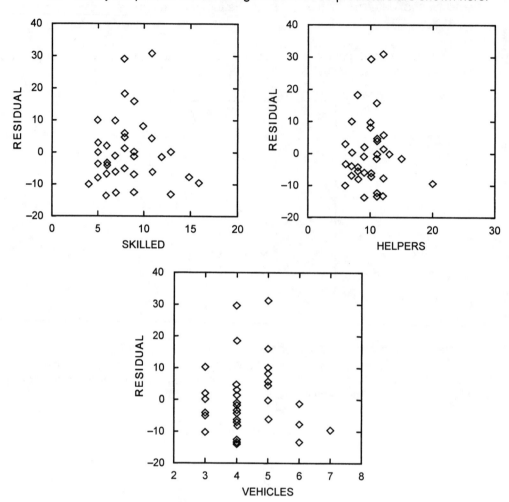

There is at least one point with rather high leverage. It shows up especially at the far right of the HELPERS plot, where one observation has HELPERS = 20, and all others have HELPERS of 15 or fewer. The Systat output in the previous exercise identified one point as high influence. This point doesn't seem to have an extremely large residual, though the equation may have been twisted toward it. The same plots should indicate whether a curved relation would be useful. We might expect a curve in this situation, because of the diminishing returns pehonomenon. Perhaps a downward-curving prediction would fit the data slightly better. There is a suggestion of a curve in the residual plots.

c. The data aren't a time series, but instead a cross section of jobs. Therefore, it doesn't make much sense to talk about the correlation of this job and the next one. The Durbin-Watson statistic is based on the relation between one residual and the next one, so it doesn't make much sense either.

14.61 **a.** We used Systat's DATA module to obtain (natural) logarithms of all four variables. A regression model involving the LOG variables was constructed in Systat's MGLH module, with the following results.

```
DEP VAR: LOGSQFT    N:  36  MULTIPLE R: 0.927  SQUARED MULTIPLE R: 0.859
ADJUSTED SQUARED MULTIPLE R: 0.846   STANDARD ERROR OF ESTIMATE:   0.097

VARIABLE   COEFFICIENT  STD ERROR   STD COEF TOLERANCE   T    P(2 TAIL)

CONSTANT       3.013      0.145     0.000 1.0000000   20.751   0.000
LOGSKILL       0.390      0.106     0.537 0.2079094    3.690   0.001
LOGHELP        0.206      0.116     0.224 0.2760822    1.774   0.086
LOGVEHIC       0.252      0.139     0.220 0.3001325    1.815   0.079

               ANALYSIS OF VARIANCE

     SOURCE   SUM-OF-SQUARES   DF  MEAN-SQUARE     F-RATIO      P

   REGRESSION       1.843       3     0.614        65.108     0.000
     RESIDUAL       0.302      32     0.009
```

The regression equation predicts the natural logarithm of 'sqft' as 3.013 + 0.390 log(skilled)
+ 0.206 log(helpers) + 0.252 log(vehicles).

b. Systat plots are shown here.

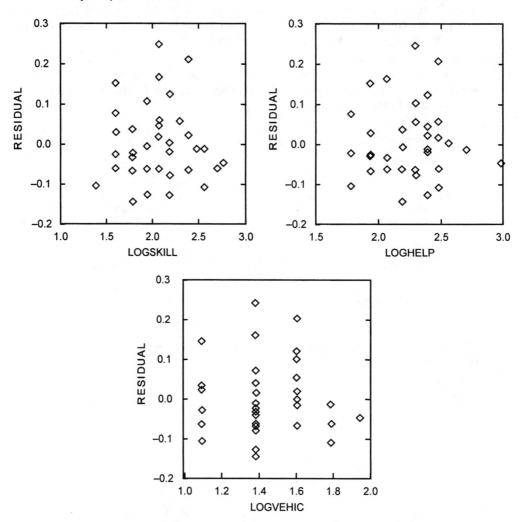

We don't see any particular curvature in any of the plots. If you do, you'll have to tell us whether the plot curves upward or curves downward!

c. We can exponentiate the entire model. As the exercise indicates, a sum of logarithmic terms corresponds to a product in the original variables. Also, a constant times a logarithm of a variable corresponds to the original variable being raised to the power. Thus the slope terms correspond to exponents of the original variable. The intercept term can be thought of as a multiple of 1; on converting back to the original variables, the intercept becomes the power of e.

$$\text{Predicted sqft} = e^{3.013}(\text{skilled})^{0.390}(\text{helpers})^{0.206}(\text{vehicles})^{0.252}$$

R.105   **a.** The equation is readable from the COEFFICIENT column.

$$\hat{y} = 1{,}160.693 + 34.562x$$

b. The slope $\hat{\beta}_1 = 34.562$ is the predicted change in total direct cost per unit change in $x =$ direct labor hours, that is, the slope is the variable direct cost per additional labor hour.

c. The intercept $\hat{\beta}_0 = 1{,}160.693$ is the estimated fixed cost (of a job with 0 direct labor hours). We note that this value is an extrapolation from the range of the data, and therefore isn't a very accurate estimate.

d. The residual standard deviation is shown in the output as STANDARD ERROR OF ESTIMATE.

$$s_\varepsilon = 3{,}462.090$$

Therefore, about 95% of the prediction errors should be within $\pm 2(3{,}462.090)$, or about $\pm 6{,}900$.

R.106   The correlation is shown in the output as MULTIPLE R (despite the fact that there is only one independent variable).

$$r_{yx} = 0.898$$

The interpretation is by the squared correlation. R^2 is shown as 0.807, meaning that variation in labor hours accounts for 80.7% of the variation in total direct cost.

R.107   **a.** The standard error of the slope is shown in the STD. ERROR column as 3.524.

b. The required t table value with $26 - 2 = 24$df is $t_{0.025} = 2.064$. $\hat{\beta}_1 = 34.462$ with a standard error of 3.524. The confidence interval is

$$\hat{\beta}_1 - t_{0.025}s_{\hat{\beta}_1}(\text{std. error}) \leq \beta_1 \leq \hat{\beta}_1 + t_{0.025}s_{\hat{\beta}_1}(\text{std. error})$$
$$34.462 - 2.064(3.524) \leq \beta_1 \leq 34.462 + 2.064(3.524)$$
$$27.19 \leq \beta_1 \leq 41.74$$

R.108 If the slope were 0, direct cost would not increase at all as direct labor hours increased. That's extremely unlikely; instead the slope should (economically speaking) be positive. We note that the confidence interval in Exercise **R.107 b.** doesn't come close to including 0, so that H_0: $\beta_1 = 0$ may be emphatically rejected. Alternatively,

$$t = 9.807$$

which is "off the t table" by a great deal, and is conclusive evidence to reject H_0.

R.109 **a.** From Exercise **R.105**, $\hat{y} = 1{,}160.693 + 34.562(890) = 31{,}921$.

b. The only change is to replace 890 with 436 in $\hat{y}$. The predicted value is

$$\hat{y} = 1{,}160.693 + 34.562(436) = 16{,}230$$

c. A job with 890 hours of direct labor is beyond the range of the data that were used to develop the regression, and is an extrapolation. A job with 436 hours is near the average for the data. The latter job will be predicted more accurately.

R.110 The description indicates that the residual standard deviation increases as x increases. Thus the constant-variance assumption appears to be wrong. The most serious effect of nonconstant variance is to make prediction intervals incorrect. Intervals centered at low predicted values will be too wide; those centered at high predicted values will be too narrow. Tests and confidence intervals are also slightly in error.

R.118 Collinearity is correlation among independent variables. We are *not* (yet) concerned about correlations between SPREAD and the other variables. We see extremely high correlations between LEADING and RATE (−0.982), LEADING and MONTH (0.997), and RATE and MONTH (−0.986). There is extremely severe collinearity.

This collinearity is also evident in the scatterplot matrix. The plot of RATE against LEADING seems to be an almost exact straight line over this time period, as do plots of LEADING against MONTH and RATE against MONTH. During the period that the data were taken, these quantities varied closely together.

Scatterplot matrices may or may not reveal nonlinear relations. They can only look at variables two at a time. Within that limitation, we can't see any curvature to speak of in any of the plots. Nor can we see any striking outliers.

R.119   **a.** To test the hypothesis that all slopes are 0, we want the overall F test. The output shows F = 321.31 and its p-value as p = 0.000. Thus the null hypothesis would be rejected at $\alpha = 0.10$, 0.05, 0.01, or even 0.001.

  b. Two-tailed p-values for the various t statistics are shown in the p column near the top of the output. The p-values for SUPPLY and RATE are small (less than 0.01) so that we may comfortably reject the respective null hypotheses that those coefficients are 0. However, for LEADING, the p-value is large; we must retain the null hypothesis that the true coefficient for LEADING is 0.

R.120   The residual standard deviation is simply denoted s in Minitab's regression output.

$$s_\varepsilon = 0.03594$$

This standard deviation is much smaller than the original SPREAD standard deviation. There appears to be good predictive value.

R.121   **a.** There is a clear cyclic pattern in the plot of residuals against time. The residuals increase for three months, generally decrease for the next 12 months, then generally increase for the remaining months. Alternatively the cycle can be seen by noticing the white spaces at the lower left, upper middle, and the lower right of the plot. This cyclic behavior indicates dependent errors, autocorrelation.

  b. The Durbin-Watson statistic is a summary statistic for autocorrelation. It is shown as 0.81, well below the ideal value of 2.00 and the rough rule for concern of 1.50. Thus this statistic also indicates that there is a clear autocorrelation problem.

  c. In the presence of (positive) autocorrelation, the residual standard deviation and the estimated standard errors of the coefficients are biased downward. They tend to underestimate the true values. Therefore, tests are too optimistic about significance. Also, R^2 tends to be an overestimate. Thus we are also too optimistic about the predictive value of the model.

R.122   **a.** By comparing the Coef columns of the difference data model and the original model, we see that the coefficients have changed very little. The coefficient of LEADING is still approximately –0.01, the coefficient of SUPPLY is slightly lower (3.35 compared to 3.75), and the coefficient of RATE is slightly lower (0.214 compared to 0.227). The partial slopes are generally quite similar. The intercept term has changed considerably, but that is of less interest.

  b. Compare the outputs for the two models (original and differenced data). Once again the overall F test and the t tests for SUPPLY and RATE are significant but the t test for LEADING is not. The results are the same as in Exercise **R.119**. Because the original output was contaminated by autocorrelation, the difference-data output is more believable.

  c. The value of the Durbin-Watson d statistic is now 2.00, equal to the ideal value. Therefore, converting to differences has eliminated the autocorrelation problem.

Differencing also affects the collinearity problem. The variance inflation factors (VIF) are now quite close to the minimum 1.0. In the original model, two of the VIF values were very high (over 28). These factors are a good indication of how severe collinearity is; the fact that they are now so small indicates that we have virtually eliminated the collinearity issue.

R.123   The value of the Durbin-Watson statistic (2.00) indicates that there shouldn't be any autocorrelation, and therefore no cyclic pattern in the data. There is no strong pattern evident in the plot, particularly as compared to the plot in Exercise **R.121**. The connected points help show the random up and down movement of the residuals.

R.124   In Exercise **R.118**, we found 3 correlations among independent variables that were larger in magnitude than 0.98. Here the largest correlation among independent variables (excluding the dependent variable DSPREAD) is less than 0.30. Thus collinearity is much lower. This result confirms what we saw with the VIF numbers, which also indicated a great reduction in collinearity.

R.145   **a.**   If the true slope were 0, there would be no (linear) relation at all between pressure difference and yield. The problem statement indicates that there is known to be an increasing relation; the null hypothesis contradicts this statement.

b.   We may either note that $F = 414.23$ with p-value $= 0.000$ or note that $t = 20.35$ with p-value $= 0.000$. However one tests, there is very strong evidence to reject H_0.

R.146 $s = 2.435$. Taking the relevant t table value as about 2 (it's actually 2.048) and ignoring the $\sqrt{1 + (1/n) + \text{extrapolation}}$ factor in the prediction standard error, we have that 95% of the prediction errors (residuals) will be within $\pm 2(2.435) = \pm 4.870$.

R.147   **a.**   The LOWESS smooth clearly curves down. That's probably the clearest way to see that we need a nonlinear model. Looking only at the points, the plot clearly has a curved shape. The residuals first increase, then decrease. This fact can be seen by noting the empty spaces in the upper left, lower middle, and upper right of the plot. (Curiously there seem to be two "stripes" of points in the plot. Perhaps a dummy variable has been omitted.) The curve in the residual plot indicates that a nonlinear model is needed.

b.   There are no residuals falling far from the main pattern of the data—that is, no outliers.

R.148   Note that in the regression with the square root of the change, $R^2 = 0.944$. This value is larger than $R^2 = 0.937$ in the original regression. Both R^2 values measure how well the original YIELD variable is predicted, so they are comparable figures. The transformed variable gives a modestly better fit (better prediction).

Alternatively: we can compare residual standard deviations, 2.297 in the transformed data model, 2.435 in the original model. The transformed model gives modestly smaller prediction errors.

Chapter 15

Time Series Analysis

15.1 Index Numbers

15.2 **a.** Using a spreadsheet, we need to sum the prices in each year to compute a simple aggregate price index. This is done below:

	Category						
	A	B	C	D	E	F	
Year 1	20.00	50.68	989.00	35,416.00	195,626.00	651,928.00	88
Year 2	18.64	48.21	1,021.00	37,215.00	206,114.00	721,200.00	96
Year 3	16.93	47.03	1,096.00	40,462.00	215,963.00	790,087.00	1,0
Year 4	16.61	46.89	1,129.00	41,943.00	229,120.00	864,326.00	1,1

Then, the index for year 2 is $965616.85/884029.68 \times 100 = 109.23$. The index for year 3 is $1047671.96/884029.68 \times 100 = 118.51$. The index for year 4 is $1136581.50/884029.68 \times 100 = 128.57$.

b. Using year 1 quantities as weights, the indices are:

$$\text{Year 2:} \quad \frac{8469222799}{8010575614} \times 100 = 105.73$$

$$\text{Year 3:} \quad \frac{9036004617}{8010575614} \times 100 = 112.80$$

$$\text{Year 4:} \quad \frac{9533701101}{8010575614} \times 100 = 119.01$$

c. Using current year quantities as weights, the indices are:

$$\text{Year 2: } \frac{8645593704}{8178083800} \times 100 = 105.72$$

$$\text{Year 3: } \frac{8992437418}{7966175280} \times 100 = 112.88$$

$$\text{Year 4: } \frac{9030079288}{7577295231} \times 100 = 119.17$$

15.3   **a.** There is a serious difference between the unweighted, simple aggregate index and the index that is weighted by yearly quantities. In the unweighted index, the numerator and denominator are dominated by the "big-ticket" items, i.e., items in categories E and F. Therefore, the unweighted index really reflects changes in these items, not in changes over all categories. When the index is weighted by yearly quantities, there is more of a balancing out across the categories. While categories A and B may have low prices (relative to categories E and F), substantially more of these items are purchased. This is reflected in the quantity weighted indices.

b. There is not much of a difference in the year 1 weighted index and the current year weighted index. This is because the combination of price × quantity did not change much from year to year within a category. For instance, while category A prices decreased from year 1 to year 4, quantity increased.

15.2   The Classic Trend, Cyclic, and Seasonal Approach

15.5   There appears to be an overall linear (increasing) trend. In addition, there also appears to be seasonal effects. For instance, there appears to be a big spike every September. In addition, within each year, sales tend to be relatively low in February, March, and April.
     There don't appear to be any cyclic effects.

15.6   The coefficient of Period indicates that every month, the overall level of sales increases by 0.596, or roughly 600 units per month.

15.7   As mentioned earlier, and confirmed by looking at the detrended data, there are some months which are consistently above and below the trend. For example, almost all February, March, April, May and June data points are below the trend (indicated by negative values). On the other hand, August, September and October are always above the trend, sometimes considerably. This suggests there are seasonal effects.

More people seem interested in purchasing car audio equipment in the late summer and fall than in the winter and spring.

15.8 **a.** The detrended value for January of year 5 is –1.63. The seasonal index value for January is –0.12. Assuming an additive model, the detrended, deseasonalized value is $-1.63 - (-0.12) = 1.51$.

b. Consider the following spreadsheet, which calculates the linear trend values from the regression equation (in the column titled "Trend"), and then adds the seasonal effects in the column titled "Trend + Seasonal".

Constant	100.513			
Period	0.59629			

Period	Month	Trend	Seasonal	Trend + Seasonal
61	January	136.89	–0.12	136.77
62	February	137.48	–11.60	125.88
63	March	138.08	–11.71	126.37
64	April	138.68	–9.53	129.15
65	May	139.27	–2.27	137.00
66	June	139.87	–9.80	130.07
67	July	140.46	2.00	142.46
68	August	141.06	13.79	154.85
69	September	141.66	21.35	163.01
70	October	142.25	9.31	151.56
71	November	142.85	–0.44	142.41
72	December	143.45	–0.98	142.47

15.9 There may be some evidence of a cyclic pattern. Consider the spike just before month 30, which is followed by 3 relatively constant data points, a jump up, and then a slow fall back down. This "pattern" is repeated at least two other times in the picture.

15.3 Smoothing Methods

15.18   **a.**

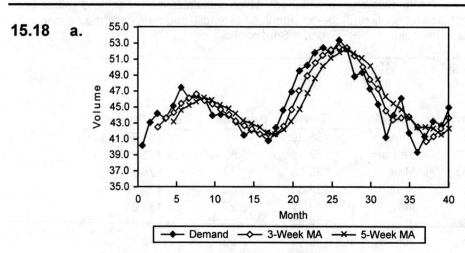

b. It appears the 3–week moving average (3–Week MA. tracks the volume time series better. It is more sensitive to the ups and downs, and does not lag behind as much as the 5–week moving average series does.

15.19   **a.**

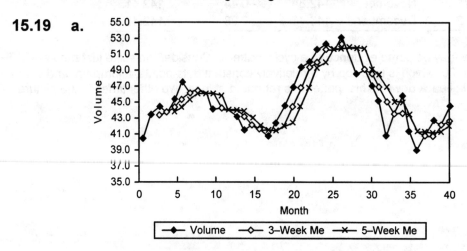

b. The better forecast seems to be the 3–week running median. It does not lag behind the movements of the volume series as much as the 5–week running median.

15.20 Using the mean squared error, the best forecasting method is the 3–week running median, with a value of 0.608. Using the mean absolute error, the best forecasting method is the 3–week running median, with a value of 0.438. Yes, both criteria give the same conclusion.

15.4 The ARIMA Approach

15.27 a.

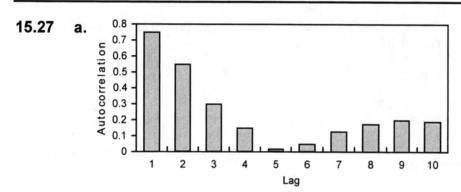

The correlation tapers off until the fourth lag, and then drops suddenly to zero.

b. In order to try and capture the taper and drop off effects, at least AR(1) and MA(1) terms should be considered.

15.28 Judging solely from a comparison of the lag correlations from the model versus the estimated correlations from the data, the more complicated models do not yield obviously better correlograms than the simpler ARIMA(1, 1) model.

15.29 Judging from the sum of squares for error, the ARIMA(1, 2) model has the lowest error, followed by the ARIMA(2, 1), and finally, the ARIMA(1, 1). Given the magnitude of the sum of squares for error terms, and the relatively small differences between them, there is not much evidence to indicate that the more complicated models fit the data much better (at least the decrease in error is probably not worth the extra complication).

Supplementary Exercises

15.38   **a.** Using the weights given, and assuming the Year 1 is the base year, the indices can be calculated by first computing the product of the price and relative quantity weights and summing. For example, for Year 1:

Year 1: $(0.66)(0.30) + (0.88)(0.28) + (0.20)(0.26) + (0.38)(0.16) = 0.5572$

Year 2: 0.6156
Year 3: 0.6904
Year 4: 0.7658
Year 5: 0.8778
Year 6: 0.9662
Year 7: 1.0764

The price index for a given year is that year's value divided by the base year (Year 1) value times 100.

Year 2: $\dfrac{0.6156}{0.5572} \times 100 = 110.48$

Year 3: 123.91
Year 4: 137.44
Year 5: 157.54
Year 6: 173.40
Year 7: 193.18

b. In a similar fashion to part **a.**, we first compute the product of quantities and prices and sum.

Year 1: 0.5746
Year 2: 0.6376
Year 3: 0.7150
Year 4: 0.7941
Year 5: 0.9108
Year 6: 1.0023
Year 7: 1.1165

Again, the price index for a given year is that year's value divided by the base year (assumed to be Year 1) value times 100.

$$\text{Year 2:} \quad \frac{0.6376}{0.5746} \times 100 = 110.96$$

Year 3: 124.49
Year 4: 138.20
Year 5: 158.51
Year 6: 174.43
Year 7: 194.31

c. It does not really matter what set of weights are used. Both choices give nearly identical results. This is because the year to year increase has applied roughly equally to all major business lines. If we were to compute simple aggregate price indices (using no quantity weights), we would get nearly the same index values. Therefore, it does not matter at all what weights are chosen, because prices in all lines of business are growing at the same rate.

15.39   **a.**

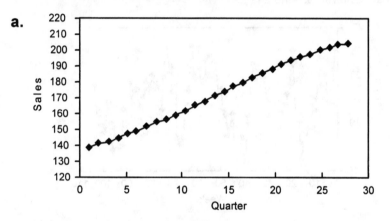

There appears to be a slight S-shape to the plot. Therefore, a logistic or Gompertz trend equation would be appropriate.

b. The linear trend equation is: Sales $= 135.1916 + 2.5599 \times$ Quarter.

15.40   **a.**

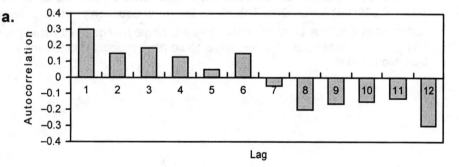

b. The autocorrelations appear to drop off sharply after the first lag.

c. Given the drop off after the first autocorrelation, an ARIMA(1, 0) (an AR(1)) would be recommended.

15.41 **a.** The AR(1) model is: $Y_t = 1.401 + 0.409Y_{t-1}$. The AR(2) model is: $Y_t = 1.205 + 0.378 \times Y_{t-1} + 0.108 \times Y_{t-2}$.

b. No, certainly not enough to warrant the more complicated AR(2) model.

c. Given the similar residual standard errors, the less complicated AR(1) model is recommended.

15.45 **a.** There is no clear evidence of a trend. In general, the series appears to be consistently centered around 21.

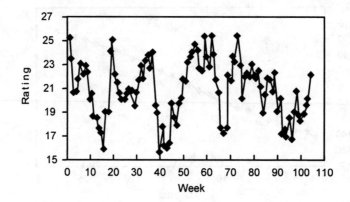

b. The trend equation is Rating $= 21.2776 - 0.008126 \times$ Week. The sign of the week coefficient indicates that the broadcast ratings are decreasing over time.

c. The week coefficient is –0.008126. This is very small, indicating a minimal trend is present, if any at all. On the other hand, taken over a year, this amounts to $52 \times (-0.008126) = -0.4226$. Therefore, the trend equation predicts a decrease of about 0.4 of a ratings point per year. Since a single ratings point usually represents many, many households, this decrease could have practical significance to the television station.

15.46 The first ten predicted values and residuals are given below.

Week	Predicted Y	Residuals
1	21.2695	3.9305
2	21.2613	2.2387
3	21.2532	−0.5532
4	21.2451	−0.4451
5	21.2369	0.5631
6	21.2288	1.8712
7	21.2207	0.9793
8	21.2126	1.6874
9	21.2044	1.1956
10	21.1963	−1.0963

For the "dog days of summer weeks", we have

Week	Predicted Y	Residuals
30	21.0338	−0.3338
31	21.0257	0.6743
32	21.0175	1.7825
33	21.0094	0.7906
34	21.0013	2.2987
82	20.6112	1.7888
83	20.6031	0.4969
84	20.5950	−1.6950
85	20.5869	−0.1869
86	20.5787	1.2213
87	20.5706	1.1294

The average of these residuals is 0.7242. This indicates that on average the model predicts less people watch during the summer than actually are watching.

15.47   **a.** The autocorrelations are given below.

Lag	Autocorrelation
1	0.74035
2	0.55594
3	0.35363
4	0.19274
5	0.02183
6	−0.03145
7	−0.0663
8	−0.0973
9	−0.09865
10	−0.1464

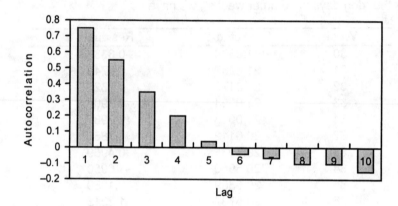

b. The autocorrelations taper off until lag 4. After this point, they drop suddenly to zero.

c. The three models and their associated residual variances are:

AR(1): $Y_t = 0.1118 + 0.7551 \times Y_{t-1}$

Residual Variance $= 2.5990$

AR(2): $Y_t = 0.1132 + 0.7440 \times Y_{t-1} + 0.0155 \times Y_{t-2}$

Residual Variance $= 2.6241$

AR(3): $Y_t = 0.1129 + 0.7451 \times Y_{t-1} + 0.1228 \times Y_{t-2} - 0.1469 \times Y_{t-3}$

$$\text{Residual Variance} = 2.5942$$

No, the more complicated models do not appear to fit the data any better than the simpler AR(1) model. All three models have relatively the same residual variance.

Chapter 16

Some Alternative Sampling Methods

16.1   Taking a Simple Random Sample

16.2   The target population is wheel-chair bound people. The sampling frame is those wheel-chair bound people who are members of the statewide association. If the statewide association of handicapped persons differs from the general population of wheel-chair bound people, then selection bias is present.

16.3   The estimated proportion is $66/150 = 0.44$. The confidence interval is:

$$0.44 \pm 1.96\sqrt{\frac{(0.44)(1-0.44)}{150}} = 0.44 \pm 0.079 = (0.361,\ 0.519)$$

Since no population size was given, the finite population correction cannot be applied.

16.6   Assuming no time trend, which would suggest changing sales over time, there is no reason to believe that this sampling method will produce any biases. Therefore, this procedure can yield a practically unbiased simple random sample. If there is a time trend (for instance a big sale which might generate higher than usual sales), this method may not yield a simple random sample.

16.7   The confidence interval is:

$$112.24 \pm 1.96\frac{91.49}{\sqrt{267}} = 112.24 \pm 10.97 = (101.27,\ 123.21)$$

16.8   If a time trend is present, autocorrelation is likely. In this circumstance, the standard deviation is likely to be smaller than it should be, causing the interval to be narrower than is should be for 95% confidence.

16.2 Stratified Random Sampling

16.11 **a.** $10.5207 \pm 1.96(0.14835) = 10.5207 \pm 0.2908 = (10.2299,\ 10.8115)$

b. $\overline{Y}_{ST} = \dfrac{(189000)(10.1) + (114000)(10.9)}{189000 + 114000} = 10.401$

$$\hat{\sigma}_{\overline{Y}_{ST}} = \frac{1}{189000 + 114000}$$

$$\times \sqrt{189000^2\left(\frac{189000 - 77}{189000 - 1}\right)\left(\frac{4.4^2}{77}\right) + 114000^2\left(\frac{114000 - 46}{114000 - 1}\right)\left(\frac{3.9^2}{46}\right)}$$

$$= 0.380228$$

$10.401 \pm 1.96(0.380228) = 10.401 \pm 0.745 = (9.656,\ 11.146)$

16.12 $\overline{Y} = \dfrac{10.4 + 11.2 + \cdots + 10.9}{8} = 10.5375$

$s^2 = \dfrac{4.2^2 + 3.8^2 + \cdots + 3.9^2}{8} = 17.745$

$\hat{\sigma}_{\overline{Y}} = \sqrt{\dfrac{17.745}{800}\left(\dfrac{1965000 - 800}{1965000 - 1}\right)} = 0.1489$

$10.5375 \pm 1.96(0.1489) = 10.5375 \pm 0.2918 = (10.2457,\ 10.8293)$.

The interval has shifted slightly higher. This is because the strata that have their weights increased tend to have higher means (such as males over 50).

16.3 Cluster Sampling

16.17 **a.** The mean is 9.567, and the standard error is 0.016.

b. $9.567 \pm 1.96\sqrt{0.016} = 9.567 \pm 0.2479 = (9.319,\ 9.815)$

c. The interval is very narrow, so, yes, the accuracy should be good enough for the managers purposes.

16.18   The mean would not be affected. The standard error would only be affected slightly. The new standard error would be 0.0164.

16.19   **a.** The individual clusters selected are denoted by the rows.

   b. $\overline{Y}_C = 114.4785$, $\hat{\sigma}_{\overline{Y}_C} = 3.855842$

   c. $114.4785 \pm 1.96\sqrt{3.855842} = 114.4785 \pm 3.8487 = (110.630, \ 118.327)$

16.4   Selecting the Sample Size

16.22 $n = \dfrac{1000(10)(40)}{(10 \times 1000 - 1)\frac{0.5^2}{1.96^2} + \frac{(1000)(10)(40)}{(10 \times 1000)}} = 579.12$

Therefore, a sample of at least 580 individuals is required.

16.23   **a.** $N = 10000$, $n = \dfrac{(10000)(80)}{(9999)\frac{0.5^2}{1.96^2} + 80} = 1094.8$

Therefore, a sample of at least 1095 individuals would be required.

   b. The required sample size is substantially reduced. This is because the estimated variability within strata is substantially smaller than the overall population variance.

Supplementary Exercises

16.27   **a.** This is a cluster sample, since all displayed dictionaries at all nearby bookstores are counted in the subset of colleges selected.

   b. This is preferred to a random sample because for the cost and effort involved, more information can be obtained. Visiting 100 bookstores selected as a random sample of all bookstores near a college would be more time consuming and expensive than visiting 100 bookstores that were associated with a smaller selection of colleges. This is because the 100 bookstores chosen as a random sample are likely to be spread out among many more colleges (perhaps even 100). With a cluster sample, a few colleges are chosen, and all bookstores nearby are visited.

16.28 **a.** Since the total number of clusters (accredited colleges) was not provided, it will be assumed to be 1000. $N = 1000$, $n = 12$, $\sum m_i = 33$, $\overline{m} = 33/12 = 2.75$, $\sum T_i = 143$, $\overline{y}_c = 143/33 = 4.3333$, $\sum(T_i - \overline{y}_c m_i)^2 = 76.2222$, $\hat{\sigma}^2_{\overline{y}_c} = \left((1000 - 12)/\left[(12)(1000)(2.75)^2\right]\right) \cdot 76.2222/11 = 0.07544$, so $\hat{\sigma}_{\overline{y}_c} = 0.27466$. The 95% CI is: $4.3333 \pm 1.96(0.27466) = 4.3333 \pm 0.53834 = (3.795, 4.872)$

b. $n = 33$, $\sum y = 143$, $\sum y^2 = 719$, $\overline{y} = 143/33 = 4.3333$, $\hat{\sigma}^2 = \left(\left[719 - 33(4.3333^2)\right]/(33 - 1)\right) = 3.104$, $\hat{\sigma}_{\overline{y}} = \sqrt{3.104}/\sqrt{33} = 0.3067$. The 95% confidence interval is:

$$4.3333 \pm 2.04(0.3067) = 4.3333 \pm 0.6263 = (3.7070, \ 4.9596).$$

(This is based on a t table value for 30 degrees of freedom.)

c. $s^2_c = \sum(T_i - \overline{y}_c m_i)^2/(n - 1) = 76.2222/11 = 6.929293$

$$n = \frac{(1000)(6.929293)}{(999)\left(\frac{0.1^2 2.75^2}{1.96^2}\right) + 6.929293} = 260.54$$

Therefore, at least 261 universities need to be sampled.

16.31 **a.** This is a stratified sampling design.

b. With a simple random sample, there is no way to assure that each order type will be represented in the sample. Furthermore, the time required to fill an order could vary substantially with order type. A simple random sample does not account for these types of difference, where they can be accounted for with a stratified random sample.

16.32 Starting anywhere in the table of random numbers, select a number. If this number is between 0001 and 8260, the corresponding order is selected. Move to the next entry in the random number table, and correspondingly select the order associated with that number. If a random number is not between 0001 and 8260, move on to the next number. Keep selecting orders in this fashion, while keeping track of how many of each type of order have been selected. Once 30 of one type of have been selected, disregard random numbers that correspond to orders of that type. In this fashion, 30 orders from each type can be randomly selected.

16.33 $\bar{y}_{ST} = \dfrac{(3.21 + 7.39 + 4.65 + 9.27)}{4} = 6.13$,

$\hat{\sigma}^2_{\bar{Y}_{ST}} = \left(\dfrac{8260 - 30}{8260 - 1}\right) \cdot \left(\dfrac{2065^2}{8260^2}\right) \dfrac{\left(0.82^2 + 2.14^2 + 1.05^2 + 3.65^2\right)}{30} = 0.04085$. The 95%

confidence interval is: $6.13 \pm 1.96\sqrt{0.04085} = 6.13 \pm 0.3961 = (5.734,\ 6.526)$

16.38 $\bar{y}_{ST} = 29.8243$, $\hat{\sigma}_{ST} = 1.80859$. The 95% confidence interval is:
$29.8243 \pm 1.96(1.80859) = 29.8243 \pm 3.5448 = (26.2795,\ 33.3691)$

16.39 Yes, the mean estimate has changed from 29.8243 to 33.7403. In the first scenario, the two smallest stratum means had the most weight, while in the second scenario, the two largest stratum means have the most weight. This explains the change.

16.42 **a.** $\bar{y}_{ST} = 182.4312$, $\hat{\sigma}_{\bar{Y}_{ST}} = 9.159808$. The 95% confidence interval is:
$182.4312 \pm 1.96(9.159808) = 182.4312 \pm 17.9532 = (164.478,\ 200.384)$

b. $N\bar{y}_{ST} = 1000(182.4312) = 182431.20$, $N\hat{\sigma}_{\bar{Y}_{ST}} = 1000(9.159808) = 9159.808$. The
95% confidence interval is:
$182431.20 \pm 1.96(9159.808) = 182431.20 \pm 17953.223 = (164477.98,\ 200384.42)$

16.43 No, because the relative sizes of the strata have not changed. The first two strata are still twice the size of the third, regardless of the overall population size.